# Fascinating Life Sciences

This interdisciplinary series brings together the most essential and captivating topics in the life sciences. They range from the plant sciences to zoology, from the microbiome to macrobiome, and from basic biology to biotechnology. The series not only highlights fascinating research; it also discusses major challenges associated with the life sciences and related disciplines and outlines future research directions. Individual volumes provide in-depth information, are richly illustrated with photographs, illustrations, and maps, and feature suggestions for further reading or glossaries where appropriate.

Interested researchers in all areas of the life sciences, as well as biology enthusiasts, will find the series' interdisciplinary focus and highly readable volumes especially appealing.

More information about this series at http://www.springer.com/series/15408

Gregor Majdic

# Soul Mate Biology

## Science of attachment and love

Springer

Gregor Majdic
Veterinary Faculty
University of Ljubljana
Ljubljana, Slovenia

ISSN 2509-6745 ISSN 2509-6753 (electronic)
Fascinating Life Sciences
ISBN 978-3-030-67211-9 ISBN 978-3-030-67212-6 (eBook)
https://doi.org/10.1007/978-3-030-67212-6



Cover photograph: Gregor Majdic Illustrations were prepared by Sandi Radovan



This Springer imprint is published by the registered company Springer Nature Switzerland AG
The registered company address is: Gewerbestrasse 11, 6330 Cham, Switzerland

*To Nada, Jakob, and Hana; hoping our love will remain true forever*

# Foreword

We have all been in love at some point in our lives. Did you do stupid things during that time, call a loved one, just breathe into the handset, and did not talk? This book gives you a rational explanation for all your irrational actions.

RTV Slovenia
Ljubljana, Slovenia

Renata Dacinger,

*The god of love lives in a state of need. It is a need. It is an urge. It is a homeostatic imbalance. Like hunger and thirst, it's almost impossible to stamp out.*

Plato

# Acknowledgments

This book was written in the last 2 years, but the seeds for it were sown a long time ago, during 30 years of my adventures in science. I have had the privilege of meeting and working with many brilliant people around the world in my quest for knowledge. These people have helped me to learn and understand science, and they have shaped my personality. Each and every one of them is a part of this book, and I remember with gratitude all the people who have stood by me through the years. Srdjan Bavdek and Rado Komel, my first mentors, introduced me to the amazing world of science. Philippa Saunders and Richard Sharp shaped my future scientific career, along with the whole gang at the MRC Reproduction Biology Unit in Edinburgh, Scotland, where we spent three unforgettable years together. Keith Parker gave me the opportunity to work in Dallas, Texas. Although I was reluctant to move to Texas, I had a wonderful time there, especially because of all the great friends in the lab. I worked with Stu Tobet for many years. He ushered me to the wonderful and curious maze of brain research. There are many other people, friends, past and present co-workers, too many to list them all, but I am grateful to each and every one of you for your support and friendship.

Special thanks go to Art Arnold, Renata Dacinger, and Barbara Radovan who read the manuscript and improved this book. I have abused the great talents of my brother-in-law, Sandi Radovan, who helped me with the illustrations.

This book would not see the light of day without the support of my family, my parents Viktor and Marija, who passed on their love of literature to me, and my beloved Nada, Jakob, and Hana.

Finally, I am extremely grateful for the privilege of living a wonderful life in a time and place free from war, famine, and other disasters. We all forget how privileged we are to live in the present time and place, while many people around the world are still suffering. We should not turn around and pretend that we can do nothing. Each of us has the chance to change a tiny part of the universe, namely ourselves. We can all change for the better. Only together can we, the citizens of this beautiful planet, build a better world for all people on Earth. A world full of love and understanding, where everyone can have their peaceful place.

# About the Book

I fell in love for the first time when I was about 4 years old. In fact, I don't even know if it's my memory or if my mother's stories have stuck in my memory. I fell immortally in love with a man at least 30 years older who worked on our terrace. From the moment he arrived in the morning, I stared at him motionlessly. While he was eating lunch, I sat on his lap. When he returned to work, I watched him. I said he was so beautiful I could not help it. And that I would marry him. The love ended when our terrace was finished.

Although I call my childhood loves love, I never thought they were true loves as we experience them later. When I read the book *Soulmate Biology*, I realized that I should be a mouse or maybe a dog if these two loves are to be real. Mice become sexually mature at 2 months, and dogs between 12 and 18 months. I also learned that both humans and elephants reach sexual maturity at the same age, between 12 and 15 years, and that orangutans are the animals that take care of the young for almost as long as we do.

After my early childhood I fell in love many times. Quite often with boys who did not return my love, and I tried my best to convince them otherwise. Sometimes by doing stupid things. I called them but I did not speak. I just wanted to hear the voice of unrequited love. Decades later, I found an excuse for my behavior in this book. The limbic system took over my actions and overwhelmed the rationality of my conscious cerebral cortex!

My last love was my husband. Reading this book, it was as if I could finally add a rational scientific explanation for our great romantic love. Although love is anything but rational, it is actually nice to know why it is like that.

After reading the *Soulmate Biology*, walking in nature or sitting by the sea is no longer the same. New knowledge makes it all the more interesting. Some male birds spend their whole life with one female and have sex with others. Which bird courts when I listen to them, and what does its wife do? Some spider ladies eat the male after sexual intercourse. What does an empty spider's web mean? Fireflies use a flashing light to court. The male hoovers and makes his light blink, and the female waits in the dark and lights up when a male she likes passes by. Male octopuses shed their penises after mating to save themselves from being eaten by the female.

Now do I know what love is? For me it is the feeling when I wake up next to my husband, spend a wonderful day with him, and go to bed with him in the evening. I may not be able to say what love is, but I can feel it. This book contains almost all the scientific answers we have about love, so I can say that I understand love better. Not completely, of course, but a little mystery is acceptable to me, and the scientists will continue their research.

RTV Slovenia
Ljubljana, Slovenia

Renata Dacinger,

# Contents

# About the Author

**Gregor Majdic** is Professor of Physiology at the University of Ljubljana Veterinary school and Professor of Physiology at the University of Maribor Medical School in Slovenia. His main scientific interest is how genes and hormones influence the development and function of the brain, and especially how brain function is related to behavior. He also studies adult stem cell treatments in animals and funded a start-up company Animacel, which provides stem cell treatments for animals in several European countries.

After graduating from the Faculty of Veterinary Medicine at the University of Ljubljana, his passion for knowledge led him to pursue a scientific career at the University of Edinburgh, Scotland, and at the University of Texas Southwestern Medical School in Dallas, Texas, before returning to his native Slovenia.

He has published more than 80 peer-reviewed scientific articles. He is a member of several international scientific societies and has lectured at many universities in Europe, the USA, and Asia. For his development of stem cell therapies in animals, he was awarded two prestigious national awards for the best innovations in the public sector in Slovenia.

He has written numerous popular science articles for various Slovenian newspapers/magazines and is a regular guest on radio and television programs about science. He is also the author of three published books, two fictional historical novels (*Gabriel's Passion* and *White Saracen*) and a popular science book for younger readers, *Secret of Life*, which describes the understanding of sex determination in various cultures throughout history.

Above all, Gregor is devoted to his family, his wife Nada, his son Jakob, and his daughter Hana. Together they love to travel around the world (on their own and off the beaten track), to cook, play, laugh, and be active in nature; they like rock climbing, mountaineering-skiing, mountain biking, kayaking, scuba diving, and other outdoor sports.

# Chapter 1
# Introduction

Love, one of the most profound of human emotions, love that follows us from puberty to old age, love that follows us from ancient times to modern, from antique writings, through the Bible, through the texts of medieval scribes to modern-day books and movies. Through the millennia love has lost none of its secrecy, charm, attractiveness, and craziness, even in this digital age, when we are overwhelmed by information.

But what is love? Where does this emotion originate? Are we humans the only living beings feeling this emotion? Can love be explained by some chemical reactions in our brains? Is love just a trick of nature or is love really some kind of higher feeling? We do not have definite answers to any of these questions; nevertheless, the fields of neuroscience, behavioral science, and others have provided us with some, at least partial, answers.

We know today a great deal more than ever before about what is happening in the brain when we are madly in love. We understand why our hearts beat faster when we see the person we love, we know why we sweat and why we feel anxious when the loved one is away from us, and we have some ideas about how feelings of attachment form in the brain.

This book will lead you through the complicated labyrinth of genes, hormones, and brain cells that are involved in the feelings of love, attachment, affection, and also simple sexual reproduction, a scientist's eye view of love, with a focus on neuroscience and behavioral science, my areas of expertise as a professor. But this is love with a twist. While we will explore humans, my research is into the emotions of animals. We, too, are animals, and understanding animal behaviors and minds offers deep insight into our own.

Love, as we humans define it, is a unique feeling originating in our large, human brain, and animals do not fall in love or feel the love in quite the same way we humans do. Yet, love has evolved as a part of the strategy for sexual reproduction, to ensure the propagation of species into the next generation. Thus, many molecular events that occur in our brains and other parts of the body when we are in love have



G. Majdic, *Soul Mate Biology*, Fascinating Life Sciences,
https://doi.org/10.1007/978-3-030-67212-6_1

their origins in the different natural instincts of animals that are broadly connected with sexual reproduction. To understand what is happening in our brains when we are in love, we must therefore know how these processes evolved and what functions they have in different animals. Hence, we will talk about different aspects of sexual reproduction and about all the processes connected with mating and rearing the young, taking a look at a variety of different animals, before we will try to explain why it happened and what is happening in our human brains when we are in love.

I do not actually want to "explain away" everything about love. I believe that part of the love must remain beyond the reach of science, beyond the shuffling of the molecules in our brains. Even if we try to understand what is happening in our brains when we are in love, love is part of our nature, our history, our legacy. It must remain a mysterious and wonderful feeling that everybody experiences, a sentiment that sows the seeds of poems into the souls of poets, emotion that brings the colors of love into the souls of painters, and this must and should stay beyond the realm of science. That said, there is much that we can learn about ourselves by learning about how the minds of animals, and our own minds, work. We do not have all the answers, nor should we, when it comes to love. But what we do know today is quite astounding and shines a brighter light into who we are.

# Chapter 2
# Two Is Better Than One

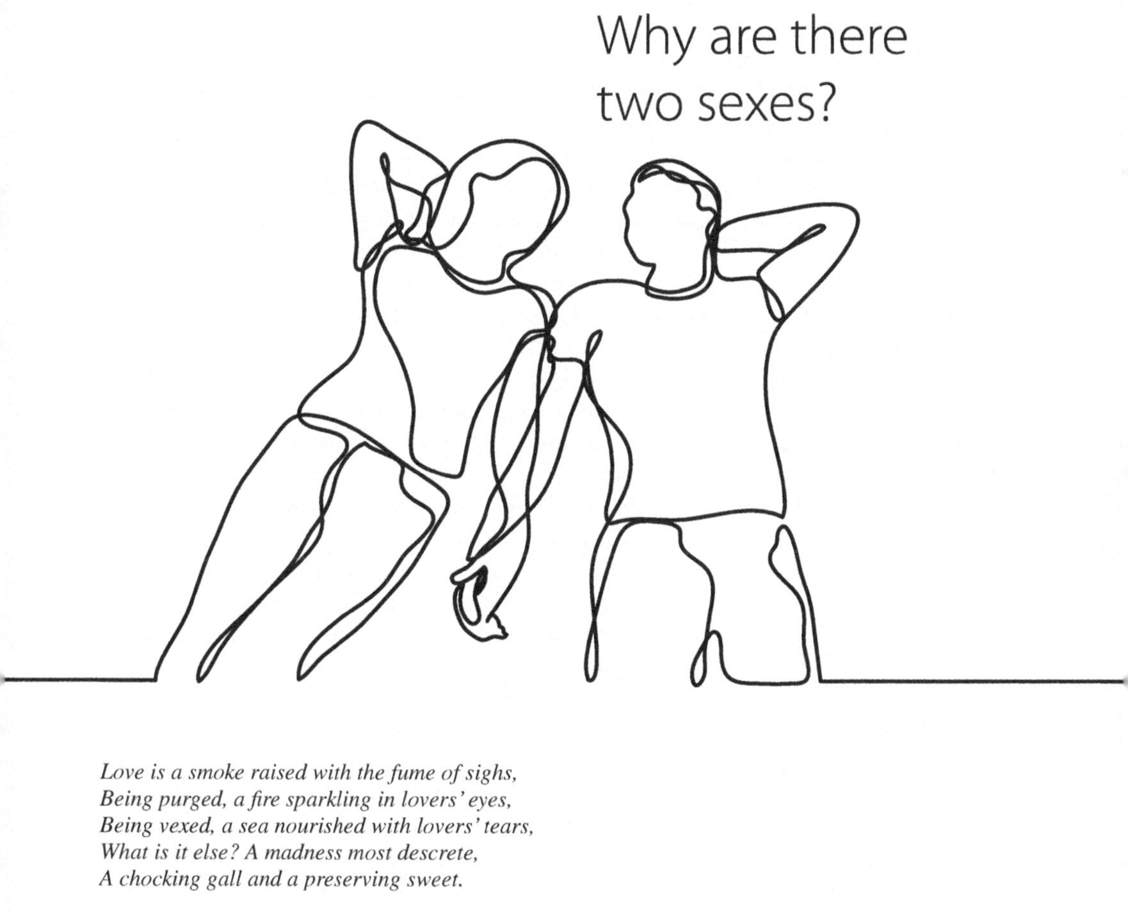


*Love is a smoke raised with the fume of sighs,*
*Being purged, a fire sparkling in lovers' eyes,*
*Being vexed, a sea nourished with lovers' tears,*
*What is it else? A madness most descrete,*
*A chocking gall and a preserving sweet.*

So wrote Shakespeare, more than 400 years ago, and although some suggest that the idea of romantic love was born in the Middle Ages with the troubadours, love, of course, has much older origins. We are part of nature, and nature has evolved over billions of years since the first living organisms appeared on earth more than 3 billion years ago.

Attachment, bonding, and, ultimately, love are more recent evolutionary inventions, but the origins of these emotions were born far deeper into the past than the

 
G. Majdic, *Soul Mate Biology*, Fascinating Life Sciences,
https://doi.org/10.1007/978-3-030-67212-6_2

Middle Ages, which are, looking through the geological history of our planet, just "few seconds" ago. Attachment, cooperation, and bonding are all evolutionary advantages, as they help individuals to survive, and the survival of the individuals is ultimately important for the preservation of the species, which is the primary purpose of all evolutionary processes. Although distinct from love or lust, the attachment of individuals to a group is seen in many animals.

Ants and bees, and some spiders, live in large colonies that work together to help their survival and spread in their environment. Fish are often seen in large schools that protect individuals and ensure the survival of the species. Pistol shrimps live together in large colonies, with individuals having their roles defined. Going further down the evolutionary tree, even bacteria often aggregate together, and the exchange of genetic material between individuals has, contrary to general beliefs, very early origins. We now know that bacteria and protozoa developed unique processes to exchange genes. This assists the survival of individuals within the species. If relying purely on clonal asexual reproduction, the species might be extinguished due to the accumulation of harmfulmutations in the genome. All these origins of cooperation ultimately lead to the development of attachment and bonding. This has eventually evolved in mammals, birds, and possibly before that in the dinosaurs into special relationships between mother and children and between male and female partners.

But why do we need sexual reproduction that ultimately, through the evolution, led to love in humans? Would it not be much simpler if we could divide in half and thus produce an offspring? Or shed some cells from which a new human would develop, just as a new tree could grow from a sapling? Indeed, asexual reproduction, present in some groups of animals, is a much simpler and much more energy-efficient process. If an organism can divide into two equal offspring, without the need to search for the partner, without the need to care for the young, it can have more time for finding resources for survival (food), more time to develop a strategy to survive in the environment, more time to evade the predators, and, therefore, much better chances of survival in the harsh world. Sounds good, does it not?

But there is a glitch. As we or any other organisms are exposed to the harmful environment, or just because our machinery duplicating our DNA is not perfect and makes mistakes, just as we make mistakes when we are typing on a computer, our genes are accumulating these errors through life. As we age, more and more such mistakes, called mutations, are present in our genes. These mistakes could eventually become harmful, leading to different diseases, including cancers, and ultimately causing the demise of the organism. If we were to reproduce by simple division, our offspring would inherit all these mutated genes and be born as if they have the same genetic age as their asexual parent, who gave them all its genes.

Therefore, it is necessary to refresh, renew, or repair the genetic material for the survival of the species, and this is achieved through sexual reproduction. Organisms that reproduce sexually have in general double (or in some cases more than double) complement of all genes, one set received from the mother and one set derived from the father. If one part of the genome is damaged, its undamaged counterpart could serve as a masterplan to repair the damaged genes.

Furthermore, sexual reproduction has another significant advantage. Every living organism must adapt to the environment to ensure the survival of the species in the ever-changing environment. Without such adaptations, animals that do not alter would sooner or later be wiped out from the environment, in accordance with the red queen hypothesis. This evolutionary hypothesis borrowed the name from Lewis Carroll's book, *Through the Looking-Glass*, where the queen of hearts once said to Alice: "Now, here, you see, it takes all the running you can do, to keep in the same place." In evolutionary terms, and very simplified, this means that every species has to evolve as fast as it can, just to keep its place in the ever-changing environment. The predatory animals are continually evolving to be more efficient in obtaining their prey. Therefore, the animals representing the prey must change equally fast as predators. Otherwise, they would face the danger of being completely obliterated by advanced predatory animals.

Without sexual reproduction, we would be clones of each other. If we would reproduce by splitting in half or if part of our bodies would bud into a new being, our children and grandchildren would possess all the same genes we have. They would look very similar, and to some extent, they would behave similarly. One might believe that this might not be such a bad idea. We would not only look the same, but we would also think similarly, behave similarly, and might, consequently, get along with each other better than we do today when we are all different.

Many science fiction writers imagined armies of clones. In the *Star Wars* saga, Darth Sidious commands the vast army of clones, all behaving similarly, with perfect order in such a uniform military. However, such an army would also be very vulnerable. As we have seen in the year 2020, during the Covid-19 pandemic, some people did not get sick at all, some suffered from very severe symptoms, and some succumbed to the disease. These differences were partially caused by the differences in the genetic make-up of each individual human being. Certainly, with a disease like Covid-19, other external factors such as smoking, being obese, and having diabetes or cardiovascular disease play important roles, but even these are, to a large extent, influenced by our genes.

Covid-19 afflicted the USS aircraft carrier Theodore Roosevelt. Out of 4500 members of the crew, more than 1000 sailors tested positive, and there was one casualty. If the whole crew would be clones of that one sailor who died, most, if not all, sailors would likely suffer seriously from the disease, and most of them would probably die. Therefore, genetic diversity is essential when we face a threat. Genetic differences ensure that the whole population does not succumb to a specific threat such as a contagious, deadly disease.

Differences in our genomes are the result of two processes. One is random mutations in the genomes that instigate new characteristics of the organism. Random mutations are, as the name tells, haphazard and thus a big gamble, a little like a lottery. One individual out of a hundred thousand or even millions might be lucky to obtain beneficial mutation(s). Chance beneficial mutations are very rare events, although in the long run, in millions of years, these random mutations are a strong drive of evolution.

More important for faster progress and especially for the survival of the species is the combination of the genomes during sexual reproduction. Sexual reproduction can provide endless combinations of genes and, thus, characteristics of the offspring in every generation. With around 25,000 genes and over three billion letters in the book of life, our genome, the number of possible combinations, is unfathomable. As Richard Dawkins said: "We are going to die, and that makes us the lucky ones. Most people are never going to die because they are never going to be born. The potential people who could have been here in my place but who will in fact never see the light of day outnumber the sand grains of Arabia. Certainly those unborn ghosts include greater poets than Keats, scientists greater than Newton. We know this because the set of possible people allowed by our DNA so massively exceeds the set of actual people. In the teeth of these stupefying odds it is you and I, in our ordinariness, that are here."

Sexual reproduction is more efficient than random mutations. It provides new, potentially beneficial novel combinations of genes. The ability to produce novel combinations of genes in every generation offers a surviving advantage for all creatures in the ever-changing environments on our planet. Some individuals will always be able to adapt to the altering environment due to different genetic make-up. These beneficial changes are what drives evolution and adaptations of organisms to their environment. Some genes are more suitable for survival in specific environments, and such genes are enriched in the populations.

In contrast, harmful genes are slowly eliminated from the populations. Darkly pigmented skin is beneficial for the people living close to the equator. However, when our ancestors moved to moderate climates with less intense sunshine in northern Europe and Asia, the dark skin was no longer beneficial, and genes for dark skin were slowly eliminated from our genomes. In the middle ages, the plague killed between 30% and 60% of Europe's population. Luckily, not all people succumbed to the disease as some were more resistant to Black Death due to different genes. Thus, our species survived in Europe, while if we would be clones, our ancestors might be wiped out completely.

From an evolutionary perspective, having two sexes is what ensures variation in the genomes and, consequently, the survival of the species. Therefore, it is not surprising that sexual reproduction developed very early during the evolution of multicellular organisms. The oldest proof of sexual reproduction is around 1.2 billion years old and was found in the fossils in Canada. Ancient fossilized seaweed, preserved in the Canadian rocks, had produced distinct male and female cells for reproduction. These fossils are, therefore, the oldest known example of sexual reproduction or perhaps, two sexes.

After its first appearance, sexual reproductions stayed on earth for good. Two sexes are present in the vast majority of multicellular organisms in both plants and animals. Interestingly, though, despite the importance of having two sexes, there is no uniform path leading to the male and female sex. Many important genetic accomplishments that occur during evolution are often preserved for a very long time. Quite a few genes retained a similar function even in such diverse organisms as fruit flies and humans. But there is no universal path to attain baby boys and baby girls.

In humans and other mammals, genes determine the sex. In every cell in our organism, DNA is organized into structures called chromosomes, which are very long chains of the DNA, coiled around the protein scaffold. We possess 46 such chromosomes in each cell. Twenty-two of them are present in each cell in two copies, in so-called homologous pairs. Members of the 23rd pair are called sex chromosomes, and these are different in boys and girls. Girls have two same chromosomes, called X due to their shape, and boys have one X and one Y chromosome (Y also called Y because of its shape).

A small gene on the Y chromosome determines the sex of the baby. If this gene, scientifically called SRY, is present in a very young unborn baby, the baby will develop as a boy. If there is no Y chromosome and no SRY gene, baby girl will be born after 9 months of pregnancy.

In birds, genes also determine the sex of the offspring, but oppositely. In birds, it is the female who is possessing two different sex chromosomes, called Z and W, and blokes have two same chromosomes, called Z.

Many reptiles attained a different approach to obtain boys and girls. Although some reptiles also use genes and chromosomes to determine the sex of the offspring, in many reptiles, the sex is regulated by the temperature at which the eggs develop. In numerous turtles, turtle boys develop predominantly if the temperature is lowish and girls if temperatures are in the higher range. In alligators, low and high temperatures during hatching induce predominantly female brood, while intermediate temperatures favor boys.

In some fishes, sex is determined by chromosomes, in others by the temperature, and in some by the amount of oxygen in the water or the fish population's density in the particular area. And not only that—many fish can switch their roles from being a bloke to be a gal, or the opposite, during their lifetime. Why evolution has changed the strategy to acquire males and females so many times is unclear. What is known for sure is that sexual reproduction is such an evolutionary success that nature will keep the concept of two sexes alive. If needed, nature will invent and reinvent novel ways to attain lads and lassies and force them to like each other, either through basic instincts or through romantic love.

The survival of the species by introducing genetic variability in every generation is the earliest origin of having two sexes, sexual reproduction, and recombination of the genes. All these, through the millions of years, ultimately led to the development of human emotions such as love, lust, attachment, and bonding. As we are not any longer driven only by instincts and our autonomous brain which would force us to reproduce, love is, at least partially, an evolutionary trick to convince us to procreate.

No matter how fruitful and beautiful our life is and should be, from a purely nature point of view, the only important thing is that we reproduce, pass our genes into the next generation, and thus ensure the survival of our species. This does not take away the meaning of life. It does not mean we should not seek and find love. We should live the years we have allocated on this beautiful planet to the fullest. We should seize the opportunity given to us by the random combination of genes from our parents. We should endeavor to build a better world for our children and grandchildren.

Sexual reproduction and continuation of our species mean that we can all achieve immortality. Neither good deeds and successes that we gained during our lifetime nor wealth and new knowledge we amassed, the most important and the longest-lasting legacy we leave on this world are the products of love, our progeny. Due to sexual reproduction and love, our genes will live long after we leave this world. Through love, we accomplish immortality, a feat that many humans tried to achieve throughout history but were unaware that most of us have already triumphed the death.

# Chapter 3
# *Escherichia coli*: Do Bacteria Get It On?

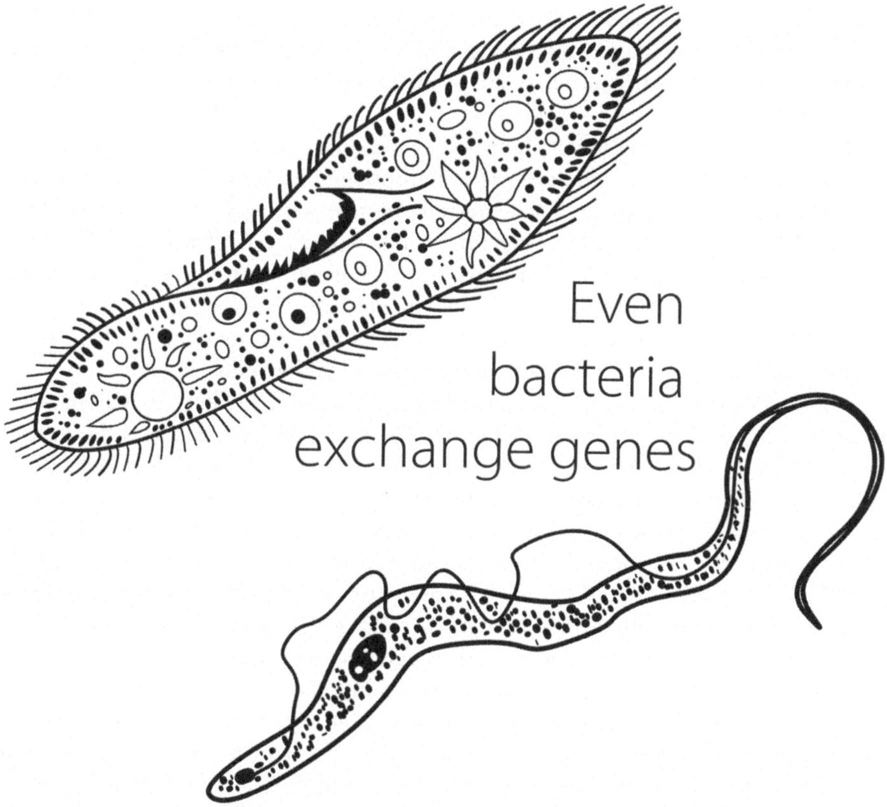


*You could also ask who's in charge. Lots of people think, well, we're humans; we're the most intelligent and accomplished species; we're in charge. Bacteria may have a different outlook: more bacteria live and work in one linear centimeter of your lower colon than all the humans who have ever lived. That's what's going on in your digestive tract right now. Are we in charge, or are we simply hosts for bacteria? It all depends on your outlook.*

Neil deGrasse Tyson: Space Chronicles—Facing the Ultimate Frontier

 
G. Majdic, *Soul Mate Biology*, Fascinating Life Sciences,
https://doi.org/10.1007/978-3-030-67212-6_3

We are surrounded by bacteria. They live on our skin and inside our intestines. We exploit them to ferment food. It is bacteria that turn milk into yogurt, cabbage into kimchi. They help us digest the food, but they also often cause a stomachache and sometimes even diarrhea and vomiting. Today we can fight them with antibiotics, but in the past, they caused many deaths. The Black Death was caused by bacteria and so was typhus. These diseases are almost forgotten today, thanks to the science that invented antibiotics.

But what do bacteria do? They are very simple organisms, and like all other organisms, they feed and reproduce and live fairly simple lives. They do not attack us or cause the disease because they are mean or evil. They just try to survive as best as they can.

If we observe bacteria under the microscope, individuals from one species are very much all alike. This similarity is not surprising, as bacteria, as we all know, reproduce by simple division. One bacterium replicates its genetic material and then divides into two equal cells. They can do this fairly quickly. *Escherichia coli*, one of the most common but usually harmless bacteria in our surroundings, can divide every 20 min. They can divide 18 times in 6 h, and with 18 divisions, we come from 1 bacterium to 262,144 identical bacteria. In less than 1 h more, with three more divisions, there will be well over two million identical bacteria.

Back in 1946, a scientist named Joshua Lederberg did not believe that all bacteria are equal. He studied intensively different species of bacteria. He was the first to show that bacteria utilize the exchange of genetic material between two cells. This exchange of genetic material is scientifically called conjugation, has been demonstrated in several species of bacteria, and earned Dr. Lederberg a Nobel Prize in 1958. In this process, one bacterium designated the male bacterium transfers its DNA into the female bacterium. Bacteria are determined to be male or female by a small piece of DNA, called F-plasmid, or sex factor. Bacteria with this small piece of DNA are labeled as males, and bacteria that do not have this factor are considered females. Of course, when we observe bacteria under the microscope, we cannot see the differences between males and females as we can see them in many animals and us, humans. And indeed, it is a far cry to call bacteria male and female as what makes somebody man and somebody woman is not only a single difference in a small piece of DNA but much more.

Nevertheless, in addition to a small piece of DNA, male bacteria have some unique characteristics. They can make a special protrusion on their surface, called F-pilus. Pilae (plural for pilus) are hair-like structures that cover the bacterial surface, but only male bacteria have these special F-pilae. These are unique structures, thicker than normal pilae. In some very broad way, we can think of them as a similar structure as a penis in mammals.

When a male bacterium meets with a female bacterium, it extends its F-pilus toward the female, and F-pilus physically connects with the bacterial wall of the female bacterium. The pilus is hollow, although very thin, and through the pilus male bacterium sends some of its DNA into the female bacterium. A female bacterium then utilizes this new DNA to repair its damaged DNA, to make a novel combination of genes and to acquire new characteristics. In essence, this new DNA helps bacteria survive in the harsh outside world, or the world inside our own organisms, which can be equally cruel for bacteria.

Our bodies are fighting the bacteria all the time with specialized cells that act as soldiers, or with particular molecules, called antibodies, produced by some white blood cells. White blood cells and antibodies recognize intruders like bacteria when they enter our body and neutralize them, just as defending soldiers will neutralize foreign soldiers entering a sovereign country. In addition to the defense produced naturally in our bodies, we are also smart enough to invent new things to help us fight bacteria, such as antibiotics.

Although bacteria are much more primitive organisms than we are, as they do not possess different organs and things like the immune system to fight something that may threaten them, they will fight back at us and our inventions. One advantage bacteria have in this battle is their quick reproductive cycle. Most bacteria divide from once every 10 min to once every 24 h. If we take 2 h for one division of bacteria (what is common in diseases causing bacteria), and thus one generation, we get 12 generations within a single day. This would, in evolutionary terms, equal close to 300 years of human lives.

During 1 week, which is usually the length of the antibiotic treatment, bacteria will go through 84 generations. This would be equal to over 2000 years of human history if we consider one human generation around 25 years. So, from an evolutionary perspective, bacteria can change in 1 week at least as much as we did in 2000 years, and in reality, even more, as their genomes are much smaller and mutations occurring randomly are more likely to cause a difference in some characteristic, either beneficial or harmful. This is a crucial weapon in bacteria—quick change and adaptation to the harsh environment.

If we take antibiotics, they will kill most of the bacteria. But due to random mutations happening all the time, some bacteria will acquire new characteristics that will help them live in the presence of antibiotics. Such resistant bacteria will divide and repopulate the niche being emptied by antibiotics. To spread this resistance to antibiotics even quicker, bacteria could utilize the exchange of genetic material—conjugation. If male bacteria acquire such new characteristics (resistance to antibiotics), they can transfer the gene for this resistance to the female bacteria, causing a rapid spread of such, initially random, beneficial mutations. This eventually creates a problem for us, as antibiotics will not work any longer.

Furthermore, such bacterial survivors of antibiotics can preserve these genetic changes for a long time. If we become ill again and start a new treatment with the same antibiotic, only a few male bacteria will be enough to resist our treatment. This time, there will be no need to wait for a random mutation that will occur (or will not) in one of the bacteria. Rather, few bacteria that already possess resistance to the antibiotic will spread this genetic information through the conjugation and make antibiotics ineffective.

However, conjugation as the exchange of genetic material between bacteria is not the only way through which bacteria can transfer genetic information between individuals. To enhance their chances of survival, bacteria utilize two additional ways to exchange their genetic material, develop new combinations of genes, and improve their chances for survival. One is called transformation and is based on the uptake of genetic material (DNA) from dead bacteria by the living ones. This uptake of the dead DNA may sound a bit morbid to us, but it is very beneficial for bacteria. When a bacterium dies, its cell wall disintegrates, and its DNA is released into the environment. As DNA is a relatively stable molecule, it can stay around intact for much longer than, for example, proteins. Bacteria that, by chance, encounter such DNA from other, dead conspecifics (member of the same species) can absorb the DNA through the cell wall. They can then use this newly acquired DNA to form novel combinations of genes that could increase their chances for survival.

The third form of exchanging genetic material is called transduction, and this exchange involves a third partner. In a very broad sense, it could be seen as a kind of artificial insemination in bacteria. Bacteriophages are viruses that attack bacteria, and after they invade a bacterial cell, they force them to produce new viruses. Viruses are composed of a protein coat and genetic material in the form of DNA (in some viruses also in the form of RNA) inside the protein coat. Once the bacterial cell has produced a large number of new viruses, such bacterium usually dies and releases the viruses in its environment. However, during the production of the viruses, quite often, parts of bacterial DNA are incorporated into newly formed viruses, alongside the viral DNA. Once such a virus infects a new bacterium and the bacterium successfully resists the attack by the virus (survive and kill the virus), the bacterial DNA incorporated into the virus genome could be used by this newly infected bacterium to obtain new genes/genetic combinations. So, although the viruses are, in principle, very harmful and deadly, bacteria can also exploit these harmful viruses for their benefit and especially for the benefit of the species.

No, the bacteria do not get it on. The exchange of genetic material between bacteria is not sexual reproduction, as the transfer does not occur as part of the reproductive process. However, this exchange of genetic material represents a very early evolutionary form of producing changes in the genomes by combining genes of two individuals. Combining genes from individuals and thus enhancing the survival of the species by providing ever-changing combinations of genes could help the species survive and thrive in the environment. As this process of combining genes proved to be highly successful already in the most primitive living organisms, in bacteria, the evolution used it, developed it, and refined it. In billions of years, it developed into many different forms in different animals, all the way to the romantic love in humans.

# Chapter 4
# Sugar-Smelling Amoebas

The origin of the smell, the most primeval sense

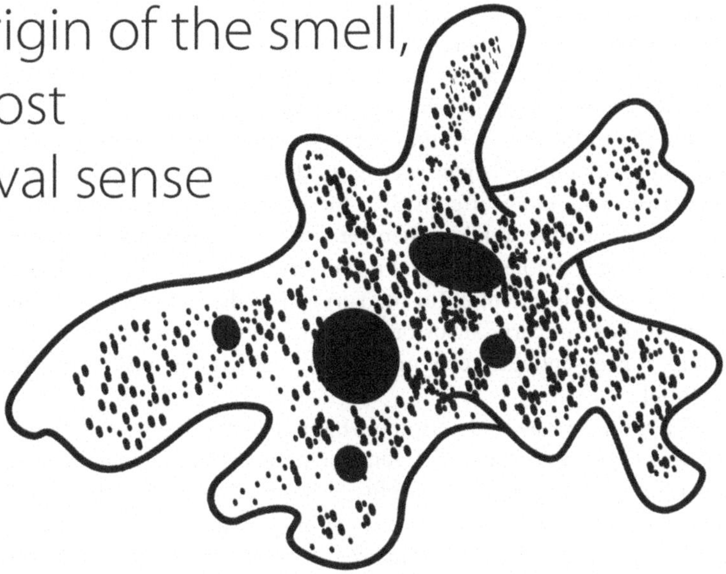

*For people could close their eyes to greatness, to horrors, to beauty, and their ears to melodies or deceiving words. But they couldn't escape scent, for scent was a brother of breath. Together with breath it entered human beings, who couldn't defend themselves against it, not if they wanted to live. And scent entered into their very core, went directly to their hearts, and decided for good and all between affection and contempt, disgust and lust, love and hate. He who ruled scent ruled the hearts of men.*

Patrick Süskind: Perfume—The Story of a Murderer

We often think that seeing is the most essential sense of love. We look at the beloved and know that their face is attractive to us. But most of us also remember the taste of our first kiss, the smell of our first love, the smooth skin of our partner when we made love for the first time. Love can be perceived by the sight of the beloved person, by hearing the voice of the loved one, by feeling our partner's soft skin, and even by tasting their skin and smelling the scent of our betrothed. We have five primary senses, and love can affect all of them. None of the five senses is unique to us, humans. Almost all vertebrates possess all five senses, although some are more



G. Majdic, *Soul Mate Biology*, Fascinating Life Sciences,
https://doi.org/10.1007/978-3-030-67212-6_4

active and important in some species and others in other animals. Humans and our closest ancestors, apes, have excellent sight. Many other mammals have poorer vision but are much better at sensing and deciphering the scents, odors, and pheromones.

All living organisms share the same two basic drives—to obtain the energy for the survival of the individual (feeding) and to reproduce to ensure the survival of the species. Two additional primal drives which are present in almost all animals (but not in plants) are fighting for survival, food, or mate and fleeing from the danger. So essentially, the survival of the individual and the species is down to four basic instincts: feeding, fighting, fleeing, and reproducing.

In order to fulfill two of these primary missions, namely, feeding and reproducing, living organisms began very early developing ways and means of performing these two tasks more effectively. One of the most important ways to ensure the survival of the individual is, therefore, to recognize what could serve as an energy source (what is edible) and what is dangerous or toxic. Recognizing what is poisonous or harmful is essential for the survival of the individual. For the survival of the species, once our early ancestors begin to reproduce sexually and stopped splitting in half, it is essential to recognize a potential mate and distinguish a partner from a potential enemy.

The simplest way to determine what is yummy and what is potentially toxic is to develop a system to detect different chemicals in the surroundings and classify them according to their potential benefits or harms. For us, smelling and tasting the food is the easiest way to tell whether something is likely to be good for us or something could cause us problems if eaten. We recognize the repugnant smell of foul meat as repulsive, and this prevents us from becoming intoxicated with spoiled meat. The smell of freshly baked bread, however, is pleasing and instinctively triggers feelings of hunger. So, although we rely mainly on seeing and hearing in our daily dealings with the world, smells remain an important part of our lives, as with most animals.

It is believed that the detection of chemicals in the environment, similar to our sense of smell and taste, is the most primordial sense developed very early during evolution. As Patrick *Süskind* wrote in his novel *Perfume*, people could close their eyes and press their ears, but they cannot escape scents. The recognition of chemicals in the surroundings is considered the most primitive of all senses. It is present already in unicellular organisms such as bacteria and protozoa.[1]

Of course, these small, one-celled living creatures have no specialized organs to perceive odors or recognize taste. They have no nervous system, let alone the brain, to analyze signals, store memories, and react to chemicals accordingly. Yet, even bacteria and protozoa possess a mechanism called chemotaxis, which is very similar to the senses of smell and taste. Bacteria and other unicellular organisms, such as amoebae, do not possess organs like the brain to analyze stimuli detected in their vicinity and determine what would be the best reaction. Yet, chemotaxis in these

[1] Protozoa—single-celled organisms that differ from bacteria by having a cell nucleus and some other organelles. Despite being single-celled organisms, protozoan cells are more similar to animal or plant cells than to bacteria. They include amoebas, flagellates, ciliates, and others.

organisms involves not only the recognition of the signal but also an immediate, autonomous response to this signal with a movement toward more pleasant chemicals or away from the harmful substances.

Bacteria or protozoa, just like most people, enjoy sugar, as it serves as an energy source. When they detect the sugar in their vicinity, they approach it and move automatically toward the highest concentration of sugar. This is called chemotaxis, and sugar is called chemoattractant, a chemical that is attractive. If, on the other hand, the compound in their vicinity is recognized as something dangerous, the chemotactic reaction is an automatic movement away from the highest concentration of such a nasty chemical. Such a hazardous compound is called chemorepellent (a repulsive chemical that repels).

Both bacteria and protozoa possess unique molecules on their cell membranes, so-called receptors, which are activated by the chemicals in their environment. The receptors are distributed all around these unicellular organisms, and the distribution of the receptors all over their cell membranes helps them to determine in which direction they should move. Although bacteria and protozoa use slightly different systems to detect the gradients of chemicals in their surroundings, the basic principle of determining the highest concentration of a particular compound is quite similar. When the receptors, which are present all around the cell membrane, detect the higher level of glucose (sugar) around one part of the cell and the lower concentration around another part of the cell, bacteria or protozoa automatically begin to move in the direction in which the higher level of chemoattractant was detected. The reverse is happening with chemorepellents, but the principle is the same. In this case, the single-cell organism will move away from the potentially harmful substances detected in one part of their cell membrane.

Chemotaxis is not only present in unicellular organisms but also inside many multicellular organisms, including our bodies. Although we as a whole organism might be attracted and even pulled toward the bakery by the smell of the freshly baked rolls, and the scent of the loved one might enamor us, this is not chemotaxis, though, in a way, it is. However, inside our bodies, some cells behave just like bacteria or protozoa in their environment. These are often cells that defend us from the enemies, so-called immune cells. Immune cells, or more commonly known as white blood cells, possess very similar mechanisms as bacteria or protozoa to move toward certain chemicals. They possess receptors for specific molecules that are released by the bacteria, and they react to these chemicals. Cells that defend us from enemies move chemotactically toward the source of chemicals from bacteria. White blood cells travel quickly toward the part of the body where the bacteria have invaded, so they can start to fight the infections as swiftly and efficiently as possible.

Other cells in our bodies also use very similar processes during embryonic and fetal development, when many cells travel from their place of birth into their rightful place in the body. Some cells travel quite formidable distances to find their proper place in the body, and this is generally regulated by chemotaxis. A great example of such migrating cells that follow chemotactic signals is many cells of the nervous system, as well as future sperm and eggs. Progenitor cells for sperm and eggs are not born inside the body of the developing fetus, but outside, in the tissue called placenta, an organ that connects a mother and a baby. These cells have to travel all the way from the placenta along the umbilical cord into the abdomen of the developing baby to finally reside in the future gonad. Although we do not yet fully understand this process, it is almost certain that it involves chemotaxis.

Similarly, brain cells are often born far away from where they will later be located. They, too, must find their correct position through the chemotactic signals. A fascinating example of such long journeys is cells that regulate our entire reproductive system and thus fertility. These cells are called GnRH neurons because they produce the gonadotropin-releasing hormone, or, shortly, GnRH, a master hormone that regulates the function of our reproductive system. GnRH neurons are born in the frontal part of the brain. They first appear in the area of the brain that will later be important for deciphering smells, directly behind the nose inside the skull. However, the proper place for these cells is in the middle of the brain, in part called the hypothalamus, quite far from the olfactory area. Chemotaxis is an essential part of this journey and guides the cells into their correct position. If something is wrong with this migration of GnRH neurons or if chemoattraction signals are missing, the consequences are devastating. One form of infertility, a disease known as Kallmann's syndrome, results from such failed migration of neural cells.

Chemotaxis is also an essential part of one of the primary drives in nature—reproduction. Have you ever wondered how the sperm find their way to the egg cell? The answer is again chemotaxis. The egg and cells surrounding it, called the cumulus oophorus, secrete certain chemicals that attract sperm over long distances. Sperm possess receptors for these chemoattractants, and when they detect them, they automatically swim toward the higher concentration. This process is particularly important in animals with external fertilization, such as fish and other marine animals, where the sperm have to find their way to the eggs through the turbid waters. Yet, it is also functional in mammals, including us, humans. Our sperm swim faster as soon as they recognize the chemicals secreted by the egg and cells surrounding it.

Chemotaxis is the most primitive and primordial of sensing mechanisms that developed during evolution. Nevertheless, it is a highly efficient system. Therefore, it has endured the evolution through the billions of years and is still present in our bodies today. Chemotaxis also gave rise to the highly specialized senses we use in our lives, such as the sense of smell and taste, and also the detection of pheromones. Chemotaxis is, therefore, rightly regarded as the forefather of our senses. Not only is it the most primitive way of how unicellular organisms communicate with their environment, but it is also the origin of some of the most important senses that animals, including humans, possess: smell and taste.

Primates developed a tridimensional color vision several million years ago. This improved vision helped our ancestors to perceive their surroundings better, and slowly the smell became less important. Nevertheless, in many mammals, the smell and detection of pheromones, with their origins in chemotaxis, are still the most important senses. Animals use scents, odors, and pheromones not only to recognize good and bad food but to orientate themselves in their environment. Scents and pheromones help them to find their way around in the forests, help animals to establish and maintain social contacts, and help them to detect intruders in their territory. Ultimately, smells and pheromones help animals to find their partners and accomplish the ultimate goal of evolution, the transfer of genes to the next generation.

Even though we now mainly use our eyes to see the gorgeousness of our sweetheart, use our hearing to listen to the lovely voice of our darling, and listen to his or her seductive words, smell has retained essential functions in our lives. Strong repulsive odors cause instinctive emotional reactions, and pleasant aromas attract us and convince us to do something that we may consciously not intend to do. Grocery stores exploit this by baking fresh bread in the shop or simply releasing the smell of freshly baked rolls into the air, triggering feelings of hunger and longing for freshly baked bread. The perfume and fragrance industries use our cravings for pleasant smells by selling us perfumes and convincing us that we should smell differently than we normally do.

Is there a role for scents in love? Do we choose our partners also by smell, not just by looks and reasoning? Do our natural scents play essential roles in shaping our relationships?

All our cells carry certain proteins, called major histocompatibility complex molecules, or MHC for short. MHCs are used in organisms to recognize our cells and distinguish them from foreign cells, including bacteria and other infectious agents that could harm our bodies. These MHC molecules are different in each person (except in identical twins), but they are more alike in genetically related individuals. In close relatives, the MHC molecules are relatively similar. More distant relatives have more different MHC molecules but still more similar than entirely unrelated people. MHC molecules are responsible for tissue rejection during organ transplants. This is the reason why organ transplantations between close relatives are generally more successful than between unrelated people.

In sexual reproduction, however, dissimilarities are encouraged as the mixing of genes leads to new genetic combinations that are beneficial for the survival of the individuals. Thus, young animal males have to leave their herds, packs, or troops and find their love and sexual partners in new, unrelated groups. For the same reason, most human societies strictly prohibit marriage and love between close relatives. Mating between close relatives is undesirable because it could empower deleterious genes, hidden in our genomes.

We all carry bad genes in us. These genes are usually not strong enough to cause harm if they are only present in one copy. So, if we get one bad copy and one good copy of a particular gene from our parents, we will be fine because a bad copy of the gene will not be strong enough to cause harm. However, if our parents are related, the probability that they both carry a copy of the bad gene is much higher. If a child inherits two copies of such a bad gene, one from each parent, he or she will become sick or be born with some anomaly. Such a disease is one of the most common genetic disorders—cystic fibrosis. One person in every 30 people carries a gene that causes cystic fibrosis. If an unrelated man and woman are married, the chances that both have a gene for the disease are small. However, suppose that father and mother are related and have a gene for cystic fibrosis present in the family. In this case, the probability is much higher that the offspring of such a couple will inherit two bad copies of the gene and develop cystic fibrosis.

Therefore, from the evolutionary point of view, it would be beneficial if animals and humans could somehow identify who has genes similar to ours. Such a person is less suitable to be our partner than someone with very different genes. MHC molecules could play just such an evolutionary role. The detection of differences in MHC molecules might influence our choice of the partner, at least to a certain extent. In mice and some primates such as gray lemur, mating is more frequent between individuals with more different MHC molecules than between animals with similar MHC genes. This prevents the mixing of related individuals and encourages mating between unrelated partners, which contributes to the greater genetic diversity in a population. Although it is not known exactly how males and females recognize the differences in MHC molecules, the recognition likely goes through the odors.

MHC molecules are proteins that are present on all our cells. When cells die and are broken down in the body, MHC molecules are excreted together with all other parts of cells, partly through the urine and sweat, which contributes to the specific body odors. Animals can recognize these scents and preferably mate with partners whose MHC molecules smell different from their own.

The detection of odors originating from MHC molecules may not be limited to our hairy ancestors. It seems that women prefer the scent of men who have more dissimilar MHC molecules to their own. Although this has not yet been confirmed conclusively and is still controversial, there is some evidence that women (more often than men) prefer the scent of genetically more different men. Such recognition of scents could sometimes lead us to decide unconsciously who we fall in love with. All this does not mean that good looks, wits, or social status do not influence our decision about our partners. They are most likely more robust drivers. We are unlikely to reject a partner because of the smell if such a partner is attractive to us because of his or her good looks, good temperament, or social status. However, if we are unsure about our choice of partner, an odor, partially derived from the MHC molecules, could decide for us whether to enter or not the long-term relationship, based on our genetic similarities or dissimilarities.

The smell could play a different role in our sexual relationships. Men seem to be more attracted to scents from women around the time of ovulation. In one such experiment, women wore T-shirts for three nights at different stages of their menstrual cycles. Afterward these T-shirts were given to the men so that they could smell them. In general, men found the scents of women around the time of ovulation more pleasant.

We are a species with concealed ovulation. Women do not send out strong, visible signals when they ovulate and are thus most fertile as most animal females do. Nevertheless, there are some indications that men can somehow discern from the smell when women are ovulating. Or, to put it the other way around, a woman might send out the signals when is the best time to fulfill our reproductive role and conceive a baby.

Several studies have examined which scents are attractive to women at different times of the menstrual cycle. Results suggest that women around the time of ovulation prefer men with more dissimilar MHC molecules. In contrast, women at other stages of menstrual cycle prefer men with MHC molecules similar to their own. This difference may seem controversial at first glance, but it can be easily explained. At the time of ovulation, a woman looks for a genetic father for her offspring. Genetically more diverse partners are preferred, since offspring from genetically different parents have better chances of survival. At all other times, when a woman seeks comfort and protection for herself and possibly a child, partners with more similar MHC molecules are preferred. These genetically related men could be relatives, who would likely take better care of closely related women to protect their family.

Now comes the interesting question—what about contraceptive pills? Pills change the levels of hormones in a woman's body, and they suppress ovulation. Do they also change partner preferences? They might actually do that. Several studies have looked women's partner preferences when they were using contraceptive pill or not. Results are usually different, which suggests that women find attractive different male scents when they are taking the pill or have stopped taking the pill. Perhaps more disturbing is the fact that women who started a relationship while on the pill might be more likely to get involved in an extramarital affair and cheat on their partner when they go off the pill, as their partner preferences change.

Nevertheless, this was only shown in one study, and even in this one the women were put in a hypothetical situation. They did not really have a chance to cheat on their partners, so the results should be taken with a large grain of salt. Contraceptive pills are certainly very beneficial for women. Their usefulness should not be questioned, but perhaps we can conclude from such studies that we need some time to really become acquainted with our partners before we commit ourselves, especially before starting a family.

Smell is probably one of the earliest senses, if not the earliest, that has its origins in the chemotaxis, firstly developed millions of years ago in unicellular organisms. In many animals, it is still the most important of all senses. It provides clues about food, but smells and pheromones also regulate social interactions and are essential for reproduction.

When we think of our loved ones, we usually recall consciously how they look like, remember their words, recall the melody of their voice. But when we think harder, we may also remember how they smell. The smells of certain people attract us, and the smell of other people repels us. Mothers recognize the smell of their children, and babies can recognize their mothers' scent at a very early age. So even if we are consciously less aware of smells than we are of sights and sounds, the smell has retained an essential function in the working of our bodies. The scents and pheromones do not have such powerful effects on us as they do on most animals, but love can certainly also be smelled, not only seen and heard. Scents might even contribute to how we choose our partners. We do not detect odors as consciously as we perceive and comprehend sights and words, but it seems that scents have retained some roles in love and romance since the early days of evolution.

# Chapter 5
# Moths in Search of Love

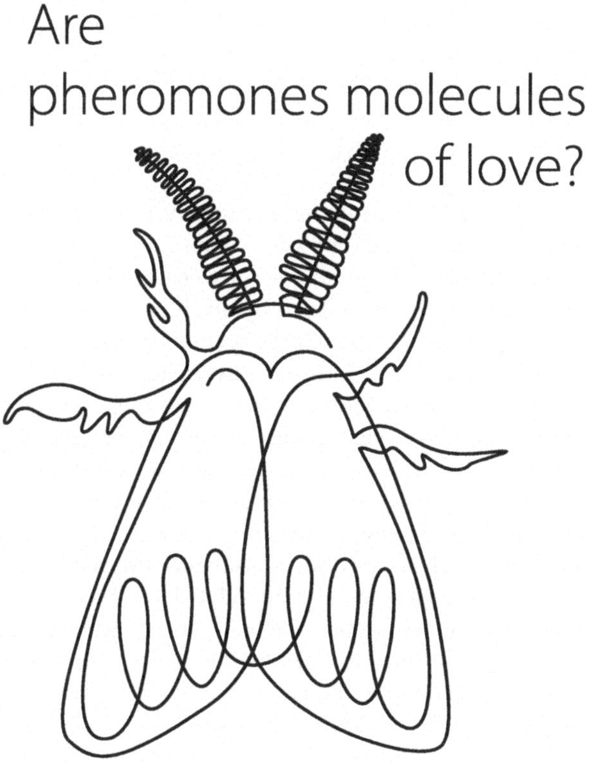


*Pheromones are earth's primordial idiom.*

Karen Joy Fowler: We Are All Completely Beside Ourselves

*Smell is important. It reminds a person of all the things he's been through; it is a sheath of memories and security.*

Tove Jansson: The Summer Book

"Pheromones for men are your secret weapon to attract gorgeous women, project powerful 'social influence,' and give you a 'high value' male vibe…" Such slogans can be read on one of many websites that sell pheromones to naive customers. If you

 
G. Majdic, *Soul Mate Biology*, Fascinating Life Sciences,
https://doi.org/10.1007/978-3-030-67212-6_5

search for "pheromones for men" on the Amazon website, you will find a selection of more than 1300 products.

Pheromones are sometimes described as molecules of love and power. If we use them, they make us instantly attractive. They will transform us into more likable people. We will be more successful at work and more popular in our social circles. If we read the reviews of such products on amazon.com, we may begin to believe that there is a magic wand that will elate our daily lives, from work to personal relationships. Unfortunately, there is no scientific evidence that pheromones or any of these products play any role in humans. It is true that in many animals, from insects to mammals, pheromones are closely linked to sexual and other instincts. However, all available evidence suggests that pheromones do not play such a role in humans.

What are pheromones? Most of us have probably heard the word pheromone, although most of us probably wouldn't know what to say when asked the simple question, "What is a pheromone?" A simple answer would be that pheromone is a chemical substance that is typically carried by air. It is produced and released into the environment by animals and affects the behavior or physiology of other animals of the same species (conspecifics). In some ways, pheromones resemble odors or taste. They excite specific specialized cells that are part of the sensory system, just as cells in the nose sense odors and papillae on the tongue recognize taste. Smell and taste give our brain information about the outside world and tell us whether something is good or bad for us. Pheromones, on the other hand, convey social information between animals belonging to the same species. In many cases, they not only pass the information. They can directly influence the behavior and even some physiological functions of organisms, usually related to reproduction or defense.

The existence of pheromones has been known since ancient times. Ancient Greek philosophers wrote about the smell emitted by bitches in heat, which was attractive to male dogs and encouraged them to mate with a bitch. In the book *The Feminine Monarchie*, published in 1623, Charles Butler wrote that the stinging bee's smell attracts other bees. Bees are summoned to join the first in stinging the victim, or rather an enemy, from the bees' point of view. At the end of the nineteenth century, an American entomologist Joseph Lintner described how the female moth attracted the male moths at his office's window. He correctly concluded that males were lured to the female by some airborne molecules. Finally, in 1959, German and Swiss scientists Peter Karlson and Martin Luscher coined the word pheromone. They derived the name pheromone from the Greek words *pherein*, which means "to carry," and *hormao*, which means "to excite." The literal meaning of the word pheromone is, therefore, "a carrier that excites." Pheromone is a fitting name, as pheromones could indeed promote arousal in many animals. They could trigger sexual arousal, aggressive behavior, or the motivation for territorial defense in various animals.

We know now that pheromones are small chemical molecules, small enough to be transmitted in the air. The first pheromone identified was a pheromone secreted by female silk moths that attract males. Five milligrams of the substance bombykol were isolated from 313,000 female moths in the 1950s. Thereafter, many other pheromones were identified in mice, hamsters, boars, and other animals.

In many ways, pheromones are very similar to smells. However, they are not the same, and, very importantly, they have very different functions. Smells are our connection to the outside world. They tell us from a distance what we can eat and what we should not touch, as it can cause stomachache. The role of pheromones is very different. In the animal world, they are a vital means of communication, one of the most important ways of interaction between animals. Through pheromones, many mammals and other animals send specific messages to other animals of the same species.

Males mark their territory with pheromones to warn intruders—do not enter the private area—just as effectively as we put up the "No Trespassing" signs on our property. Most of us have seen male dogs often urinating during the stroll. They don't have problems with their prostate or urinary system; dogs mark their territory with pheromones. Have you ever experienced the cat rubbing its head against you? It marks you with pheromones produced by special glands located on different parts of the cat's head. Have you observed a cat scratching various objects with its front paws and claws? It doesn't do this to annoy you and make you angry because it is ruining your new, expensive sofa. It is just marking the territory with scent glands between its toes.

The wilder, larger relatives of our pussycats show similar behavior. Lions, leopards, and other big cats often rub their heads together, show their affection, and mark each other with pheromones. They also often scratch trees with their paws to mark their territory with pheromones. Many animals produce pheromones to warn other members of the species of the impending danger. Such pheromones have been identified in mice. When mice face the looming threat, they freeze—stand completely still and quiet to deceive predators, who often rely on the detection of movement during their hunting sprees. At the same time, mice exposed to a threat secrete pheromones. These pheromones cause other mice in the vicinity to freeze immediately and automatically, without perceiving a predator.

Besides marking the territory and claiming property, pheromones are essential for reproduction. In contrast to humans, animals, except bonobos, do not have sex for pleasure but only for reproduction. Females are only ready for mating at the time of ovulation. To ensure that potential partners know when they are ready for mating, females secrete pheromones during the heat or estrus phase to invite the males to mate. Hamster ladies place the pheromonal scents along the path to their burrows on the night before entering the estrus. Thus, they provide road signs for males to find them. Pheromones secreted by the females in the estrous serve as an invitation for males to approach the females in many animals. They often also excite males, motivating them to mate.

Odors and pheromones are the most ancient of all senses. Even protozoa use the detection of chemical molecules in their surroundings to evaluate close environment. Amoebas detect chemicals in their surroundings and swim toward or away from the source of such compounds, depending on the message the molecule conveys. Pheromones are part of this ancient system of detecting chemicals. Animals detecting the pheromones are often instinctively attracted to the source of pheromones, just as an amoeba is drawn to the higher concentration of sugar. Yet, pheromones are not used to identify clues from the environment but evolved for the communication between members of one species (interspecies communication). Pheromones are usually species-specific and convey the messages only to the members of the same species. They can be seen as paper letters we used to write to our friends or enemies. Love letters or threatening letters, or emails and text messages we use today, in the time when we forgot that people used to write paper letters to their loved ones. Most animals use pheromones for conveying various social communique. The probable exceptions to this widespread use of pheromones are higher primates, including us, humans, where pheromones were likely replaced by other means of communication.

Although we mostly talk about mammals, many other animals use pheromones for communication. Ants are applying pheromones on the ground to mark their paths. Ant queens use pheromones to tell all the workers in the colony she is the one and only queen and workers should obey her. Bees use pheromones to tell other bees about the danger; moth (and likely other insects) females use pheromones to attract potential males for mating. Bark beetles use pheromones to communicate between themselves when they are attacking a new tree.

Mammals use pheromones in the contexts of sexual behavior, territorial protection, and dangerous situations. Pheromones secreted by a mare in heat trigger lust and craving to mate in a stallion. Female Asian elephant secretes pheromones by urine when in heat, telling the males she is ready to reproduce. Boars are secreting pheromones in their saliva. Sexually aroused boars are chomping with their jaws and secreting large amounts of saliva with, for us, a repugnant odor. In the vicinity of sows, boar snorts and spreads the spit over the sows' head. If in heat, the sow responds to the pheromones in the boar's saliva and adopts a unique mating stance, called "stand."

Boar pheromones were chemically identified. Today, pig farmers can buy boar pheromones in a spray and use them to detect sows in heat without the need for a boar. Although the word pheromone on a can that looks like a deodorant might seem tempting, it is probably not a good idea to buy it and use it for personal use. Boars' pheromones are certainly not attractive to us and will not increase our romantic chances. On the other hand, if one would like to have plenty of room in a bar, on a crowded public bus, or on the London tube during the rush hours, boar pheromones for deodorant might be the right solution.

Boar pheromones are not present only in the boar saliva but also in a costly delicacy—truffles. The best way to find truffles, special mushrooms growing under the ground, is to use pigs. Truffles contain tiny amounts of the very same chemical that serves as a pheromone in boars' saliva, and pigs are therefore attracted to their smell. Luckily, truffles contain so little of this smelly chemical that truffles still smell pleasant to us. Yet, there is enough of this chemical to be detected by pigs. The only problem with pigs as truffle hunters is that they like truffles just as much as we do. So, with pigs, it is always a fight who will get the truffle: the finder, the pig, or his owner, a human.

The presence of pigs' pheromones in truffles is not the only known example of the use of the same substance in different creatures. Some pheromones overlap between different organisms. Asian elephant cows secrete certain pheromones when they are in heat. This same substance is very attractive to some butterflies. So, if you see a large number of butterflies hovering around the backside of the female Asian elephant, this could be a consequence of natural confusion with the pheromones. Bark beetles use pheromones that are very similar to the aggression-promoting pheromones in mice. Catnip is a plant so loved by cats that it is a challenge to grow it in a garden together with cats. Cats constantly roll in catnip and destroy it because this plant secretes a pheromone that is attractive to cats. Catnip triggers behavior in cats that resembles feline mating rituals. Of course, catnip does not emit this pheromone to attract cats, which would obviously be counterproductive. The secretion of the pheromone is most likely part of the catnip's defense against aphids which love its soft leaves, and it only coincidentally attracts felines.

Some animals developed dirty tricks in the use of pheromones. When fire ants fight other ants, they release a defense pheromone, calling other ants from the colony to join the fight. However, if fire ants are unfortunate, ant-decapitating flies are snooping on the pheromones and attack the smelling ants. These flies take their name from the fact that they lay their eggs in the ants' heads, and when larvae develop in the ants' heads, the heads eventually drop off, and the ants die. So, by calling for help in the battle with pheromones, these ants bring death and destruction upon themselves.

American bolas spiders produce the same pheromones that are used by female moths to hunt their prey. When a male moth detects these pheromones, it is understandably deceived and attracted to the source of pheromones. But instead of reaching the ecstasy of love in the embrace of the female moth, it finds the death and demise at the source of the pheromone, a hungry bolas spider.

Pheromones are potent molecules, and they can have profound effects on the body. With the exception of humans and some other primates, mammals have a special organ, the vomeronasal organ or Jacobson's organ, to detect pheromones. This paired tubular organ is located between the nose and mouth. Interestingly, and contrary to popular belief, although pheromones are airborne molecules similar to odors, most mammals detect pheromones through their mouths rather than through the nose. In most mammals, the entrance to the vomeronasal or Jacobson's organ is in the front part of the upper jaw, right behind the upper lip. Most of us probably remember seeing a dog on a walk sniffing intensely, lifting the upper lip, or perhaps watching the stallion before mating, again lifting the upper lip. Domestic cats, lions, zebras, and many other male mammals show similar behavior which is called flehmen. The role of lip lifting is to open the doorways into the vomeronasal organs, so that more pheromones can enter the organ.

Once inside the vomeronasal organs, the pheromones are dissolved in the mucus that fills the vomeronasal tubes. Only then can pheromones bind to specific receptors, proteins on the surface of the cells that accurately recognize pheromones. Once activated by the binding of the pheromones, these sensory cells send their signals to the brain region called the accessory olfactory bulb. The accessory olfactory bulb is located near the area responsible for receiving olfactory signals from the nose, the olfactory bulb. However, after this initial entrance into the brain, the pathways for odors and pheromones in the brain are mostly different. Messages about smells are partially deciphered already in the olfactory bulb. They are then transmitted to other parts of the brain such as the hippocampus (site of the memory), the cerebral cortex (our conscious brain), and the amygdala (emotional part of the brain). Odors in these parts of the brain can trigger strong reactions such as fear, disgust, repulse, and even vomiting if the smell is unpleasant.

Although some of the signals triggered by pheromones also travel to the amygdala, most of the active neural connections are from the vomeronasal organ to the part of the brain known as the hypothalamus. This inner brain region, located at the bottom of the brain, near the skull's bed, plays a unique role in regulating many autonomous functions in our bodies. These include regulation of the body temperature, appetite, thirst, and internal clock, all of which are important for the survival of the individual. Additionally, the hypothalamus is also responsible for the survival of the species. It regulates the function of the reproductive organs, the production of eggs and sperm. It also has crucial roles in regulating behaviors that are important for reproduction and social behaviors. Both aggressive and sexual behaviors have their origins in the hypothalamus, and pheromones regulate both in most mammals.

Pheromones are often used by males to mark their territory, and males often leave them as a warning for potential territorial intruders of the same species. These pheromones are best known in mice but are also present in many other animals such as large cats, wolves, and others. African antelopes red and gray duikers secrete pheromones from glands around their large, beautiful eyes. They smear them on the trees and bushes within their territories to ward off potential intruders.

Territorial pheromones often act directly in the hypothalamus and trigger an aggressive response, forcing males to engage in combat to protect their territories or to invade the new lands. Some of the aggression-regulating areas in the hypothalamus have been identified very precisely. The electrical stimulation of such regions, which normally respond to pheromones, triggers aggressive behavior in male mice in the absence of pheromones. With the activated area for defensive aggression, such male mouse attacks anything in the vicinity: a plush toy, an investigator's hand, a large chunk of food, and other objects.

Pheromones secreted by estrous females arouse males of many species instinctively. They trigger the onset of male sexual behavior rituals in males. Similarly, pheromones secreted by males willing to mate induce the behavioral response in receptive females. Signals started initially in the vomeronasal organ by pheromones travel through the accessory olfactory bulb into the hypothalamus. In the hypothalamus, these signals generate a strong desire for mating and instinctively initiate behavior patterns that enable coitus.

These two examples are perhaps the most obvious and well-known pheromonal effects. However, different pheromones have many other effects, some of which are very striking.

There is always a competition in the animal world who will pass on their genes to the next generations. Many males, especially big cats like lions, and primates like chimpanzees, sometimes kill the babies that are not their offspring.

In mice, this infanticide has evolved even further, and a male mouse does not have to resort to physical violence. If a mouse lady in the early stage of the pregnancy is housed together with a male who had not sired her pups for up to 4 days after mating, the pheromones from this new male partner trigger an abortion in the female. Hypothalamus is crucial for triggering sexual behavior and sustaining early pregnancy by regulating hormone secretion from the pituitary gland and ovaries. Pheromones secreted by the male interrupt this hormonal network essential during the early stages of the pregnancy. Pups, although already present in the womb of the female, will not implant into the uterine wall. They are ejected from the uterus due to increased cramping of the uterine wall. This abortion triggers the onset of the female's estrous after only a few days, and a novel male partner is consequently able to mate and produce new offspring, carrying his genes.

Other effects of pheromones are not as drastic but could be equally crucial for some reproductive functions. For example, if young rodent females around puberty are placed together with an older male, this male's pheromones stimulate pubertal development. Such females enter puberty earlier than their conspecifics, growing in the absence of an older male. In rabbits, lactating female's pheromones induce suckling reflex in younglings, instinctively stimulating them to seek the nipple and nurse. Interestingly, this particular effect is not regulated through the vomeronasal organ but through the nose and the regular olfactory pathway. Why is this so is not known, but perhaps the vomeronasal organ is not yet sufficiently developed in the newborn bunnies to support such crucial instinctive behavior.

An intriguing effect of pheromones is the synchronization of the estrous cycle. Female mice housed together and exposed to the male mouse pheromones synchronize their estrous cycles. Most likely with the aim that females will enter the heat more quickly and the male who emits pheromones could impregnate them sooner.

A similar effect, synchronizing the menstrual cycle, has been attributed to the human pheromones. A widely cited study published in 1971 reported that women synchronize their menstrual cycles when they live together. A follow-up of this study suggested that pheromones, released through women's armpits, are a culprit of such synchronization. However, these studies have been widely criticized. Numerous subsequent studies could not confirm neither the synchronization of the menstrual cycle in women nor the role of pheromones.

Not only in the synchronization of the menstrual cycle, but there is also no evidence that humans use pheromones at all. Despite advertising of dubious companies that sell so-called pheromones to naïve customers, there is no proof that humans use pheromones for any communications between the members of our species. There is certainly no evidence that any such molecules could make us more attractive to members of the opposite sex. On the contrary, there are several persuasive proofs that pheromones, if exist at all, play a much less important role in humans than in other animals.

We do not have the vomeronasal organ, which is used to detect pheromones in animals. The vomeronasal organ initially develops in the human fetuses, but it regresses during development. Adult humans have only rudimentary remains of this organ, or often not even that. Nor do we have an accessory olfactory bulb. This absence of accessory olfactory bulb is not a conclusive proof that we cannot detect pheromones. Some argue that we have specialized cells to detect pheromones within our main olfactory bulbs, which is important for the decoding of the olfactory signals. However, the absence of both the vomeronasal organ and the accessory olfactory bulb strongly suggests that pheromones do not play an important role in our lives.

The third assumption that pheromones are not necessary for our communications comes from the fact that some genes for pheromone receptors are no longer active in humans and some primates. Due to mutations accumulated in pheromone-detecting genes during evolution, these genes are no longer functional. They become so-called pseudogenes, genes that are present in our genomes but have lost their function during evolution. Evolutionary genetic studies have shown that the gene TRPC2, coding for a protein necessary for pheromone signaling in the vomeronasal organ, developed into the functionless pseudogene several million years ago. The loss of the functional TRPC2 gene occurred at about the same time as three-dimensional color vision developed in primates.

These two coincidental changes suggest that much better-developed vision has likely replaced the role of pheromones in higher primates, including us. This fits well with the visual signs that many primate females produce during estrus such as an enlarged red vulva in female chimpanzees. From an evolutionary perspective, this makes perfect sense. Although pheromones are detected over longer distances than visual signals, visual signals are more accurate regarding the source of the message. If a male of certain species detects pheromones emitted by a female of the same species in heat, it could struggle to find the right female who emitted the signals. Furthermore, different species may use the same chemicals as pheromones for various purposes, as previously described for elephants and butterflies and bark beetles and mice. This use of the same pheromones could cause some confusion in the animal world. However, if a male chimpanzee sees a female with an enlarged red vulva, he knows that this is the female willing to mate.

Pheromones are certainly not molecules of love. There is no conclusive proof that pheromones have any physiological functions in humans at all, lest that they could trigger the emotions of attraction in the opposite sex. Therefore, a simple drop of so-called pheromones will not trigger feelings of love or lust in a member of the opposite sex, despite all the slogans and testimonials on the websites of companies that sell pheromones. Although the "pheromones" can be purchased over the Internet, they will not cause anyone to fall in love with us, regardless of how much of these pheromones we apply to our bodies or clothes.

Even if we will establish a role for pheromones in humans, they are certainly not involved in feelings of love, bonding, and attachment, as even in animals, pheromones do not have such functions. Pheromones in animals regulate lust and desire to mate, and aggressive defense of one's territory. They trigger and regulate instinct behaviors associated with reproduction, but not with attachment and bonding, which are some of the most important components of love. Feelings of love and devotion are dependent on other brain pathways that are not directly connected to pheromones and the vomeronasal organ. Therefore, the pheromones are not molecules of love. Even if we will eventually discover that pheromones play certain roles in humans, their role might be in one-night stands, but certainly not in promoting or sustaining long-term, lovable relationships.

# Chapter 6
# Wild Bull and Placid Ox

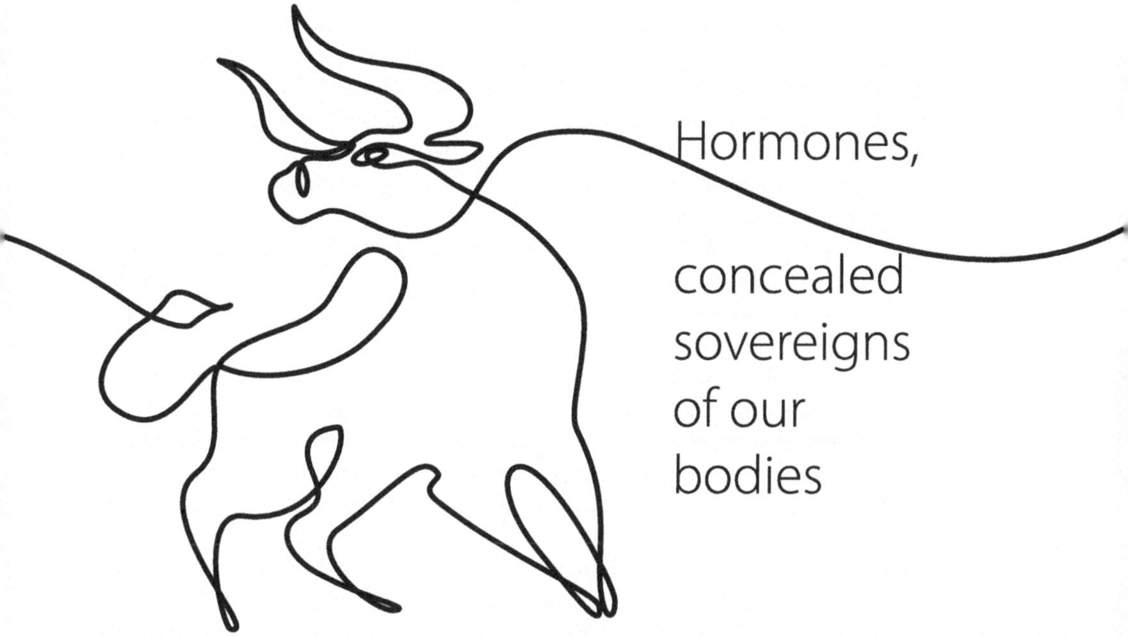


*Stud males might be emotional, temperamental, and developmentally stunted, at the mercy of their androgens, but that didn't make them incapable of generosity, friendship, cleverness, or creativity.*

Elizabeth Bear: Carnival

*And anyway, never, never apologise the morning after for what your hormones were telling you last night.*

Paul Russell: The Coming Storm

About 2000 years ago, Saint Paul wrote in his first letter to the Corinthians: "The women should keep silent in the churches." In the Middle Ages, these words were taken literally. This has a very long legacy in gender inequality that unfortunately persists in the present days. In practice, this meant that women were also not allowed to sing in the church choirs. In the Renaissance, when people began to appreciate arts more than through the Middle Ages, some clergymen missed sonorous women's voices inside the churches. Pope Clement VIII solved this problem in 1599 in a



G. Majdic, *Soul Mate Biology*, Fascinating Life Sciences,
https://doi.org/10.1007/978-3-030-67212-6_6

very inhumane way, by admitting men, emasculated before puberty, into the papal choir.

At that time, the practice of castrating prepubertal boys (usually around 8 years of age) was at least 50 years old, as the first written notes about so-called "castrati" singers originate from the middle of the sixteenth century. The popularity of male singers with extraordinary musical abilities increased in the seventeenth century, when Pope Innocent XI banned women also from singing on the stages in ever more popular opera houses. Thus, these mutilated men replaced women performers, and throughout the sixteenth, seventeenth, and eighteenth centuries, castrati, or virtuosi, as they were called more often in Italy, were hugely popular. Some of them, like Carlo Broschi, who took the artistic name Farinelli, had a similar status as today's rock stars. According to the legend, a woman, listening to his concert in London, screamed "one God, one Farinelli."

It is hard to imagine today how widespread barbaric behavior of castrating young boys was in those times. The mutilation of young boys was officially banned. Nevertheless, many parents did it. They hoped that their sons would achieve similar status as Farinelli and other stars of that time and thus bring wealth to their families. As the procedure of castrating young boys was officially prohibited also by the church, the parents did it in secret and provided various excuses. For example, it was claimed that Farinelli fell from the horse and landed so unfortunately that he lost his testicles. Other boys had experienced similar falls or had even stranger accidents. Wild boars attacked some of the boys, and these wild boars deliberately mutilated their privates.

Since medicine was not as developed then as it is today, such an operation, often performed by local barbers, was a hazardous procedure. Many boys died as a consequence of the mutilation. Even of those who survived the operation, only a few became rich and famous. This did not stop the parents from looking for the gold pot at the end of the rainbow, even by sacrificing their children.

At that time, people did not know what happens in the body when male gonads are removed. Nevertheless, they observed that when these boys grew up, they were usually very tall and had long limbs, more feminine features, no facial hair, and no Adam's apple, and, most importantly, they sang beautifully. They had higher voices than ordinary men. They were able to produce a vast range of sounds, and, due to their large ribcages, they could sing very long without pausing for breath. Some virtuosi could apparently sing phrases up to 60 s long!

A lack of hormones causes all these effects, more specifically, lack of male sex hormone testosterone, which is produced by the testes and transported by the blood throughout the male body. However, in the time of castrati, not even the word hormone existed.

The British scientist Ernest Henry Starling was the first to use the word hormone. In 1905, he delivered a lecture at the Royal College of Physicians. He reported about the discovery of the substance that is made by cells in the small intestine that stimulates secretion from the pancreas. He was the first person in the world to use the word hormones in his lecture, saying: "These chemical messengers, however, or "hormones," (from Greek word 'hormao' which means to excite or arouse), as we might call them, have to be carried from the organ where they are produced to the organ which they affect by means of the bloodstream and the continually recurring physiological needs of the organism must determine their repeated production and circulation through the body." Starling's discovery quickly led to the new branch of medicine, called endocrinology, which specifically deals with the effects of hormones in our bodies. In the years after Starling's discovery, many new hormones were soon discovered and their effects explained.

The existence of substances that travel through the body and trigger different effects has been known for much longer. The famous Greek physician Hippocrates explained epilepsy as the transport of phlegm and bile into the brain from the rest of the body through the veins. Although we know today that he erred, his notion that substances can be produced in one part of the body and transported by the blood to another part of the body was important for understanding the workings of our body.

In roughly the same period when Hippocrates lived, Chinese scholars experimented with bodily substances. In Chinese medicine, urine was considered part of the blood, and as such, the Chinese believed that urine has at least some of the same properties as blood. They experimented with extracting the contents of blood from the urine, and they developed sophisticated methods for isolating sexual hormones from the urine. These hormones obtained from urine were used in very similar ways as hormones are used in modern medicine today. Chinese doctors used an extract from the male urine to treat male infertility, low libido, erectile dysfunction, and lack of facial hair in men. Extracts from female urine were used similarly for treating problems with periods in women and women's infertility.

The effects of sex hormones on animals have also been known for a very long time. Historical records show that the castration of domestic animals was already used 8000 or 9000 years B.C. In ancient times animal breeders learned that removing the testicles from wild bulls turns them into more placid oxen. Castration of boars eliminates the unpleasant odor of the pork meat. Castration of stallions makes them easier to handle and ride. Although more difficult as they have testicles inside the abdomen, castration of cockerels has also been performed for a long time to produce animals with tastier, more tender, and moist meat. Most of these practices are still used in modern time farming and animal breeding.

Today, we know and understand that hormones play crucial roles in the functioning of our bodies and animal bodies. Hormones and the nervous system act together to control all functions of our bodies. The significant difference between these two systems is that the actions of the nervous system are almost instantaneous, while hormonal regulation is somewhat slower. The hormone has to be released from the particular cell, must be transported by the blood to the other part of the body, and can trigger the response only when it reaches the target cells. The field of endocrinology, the science about hormones, has become increasingly important in recent decades. We have learned that not only the classical glands such as the pituitary gland, thyroid, pancreas, adrenal glands, and gonads produce hormones but that just about every organ and tissue in the body produces some hormones, which are not necessarily directly related to the primary function of that organ.

For example, most of us have heard about the erythropoietin, a hormone often illicitly used by athletes to improve their performance. Erythropoietin increases the production of red blood cells whose primary function is to supply oxygen to all cells in the organism. This increase in red blood cells helps athletes to perform better because more red blood cells mean more oxygen for muscles. Interestingly, this hormone is produced in the kidney, an organ we do not usually associate with hormones, as its primary function is to eliminate harmful substances from the blood through the production of urine.

For a long time, fatty tissue has been considered to be a very inactive tissue that merely serves as an energy storage. Today we know that fat cells are not only the storeroom for fat. They are very active endocrine cells that produce several different hormones. Some of these hormones are involved in the regulation of body weight and are therefore related to the function of the adipose tissue. Some others have very different functions, such as the female sex hormone estradiol, whose primary role is very different from the function of the fat tissue. However, it is also produced in fat cells.

So, what are hormones? They are substances carried by the blood through the body that influence the function of other cells and tissues. Some hormones affect only particular types of cells. In contrast, others influence the whole body and almost every tissue or organ. Some hormones act on cells in a completely different part of the body (what is called an endocrine effect). Others affect only cells in their vicinity without entering the blood (paracrine effects). Chemically, hormones are different substances. Some are large proteins like the well-known insulin, and others are small proteins, made up of a few amino acids, like oxytocin. Some are very similar in structure to cholesterol (stress hormones and sex hormones), and some are modified amino acids like the famous adrenaline.

For the proper functioning of the endocrine system, cells that produce hormones are not sufficient. There must also be cells that are willing to listen to these chemical messengers and understand messages, cells that possess hormone receptors. If we want to talk to someone on the telephone, the person we want to talk to must also have a telephone. Otherwise, we might speak to our phone, but our words will be lost.

Similarly, a cell that produces hormones must have recipient cells somewhere for hormones to be effective. Only cells with appropriate tools, hormonal receptors, can receive the signals and listen to them. Hormone receptors can be found on the surface of the cells or inside the cells. These receptors enable cells to respond to hormones and perform the function the hormone wants them to do.

Skin cells on the face begin to grow a beard when they receive such instructions from the male sex hormone testosterone. In women, cells on the chest begin to multiply and develop into breasts if stimulated by the female sex hormone estradiol. Bones grow if instructed to do so by growth hormone, and cells throughout the body store sugar if they are stimulated by insulin.

Many hormones also enter the brain and could influence the function of our central processing unit. The male sex hormone testosterone increases aggressive behavior. The stress hormone cortisol prepares our brain and other parts of the body to face the danger, fight for survival, or flee from the threat. Leptin, secreted by our fat cells, tells the brain how full our energy stores are and accordingly triggers feelings of hunger or satiety.

Without hormones, our bodies do not function properly. Probably the best-known disease that is caused by the malfunction of the hormonal system is diabetes, a condition causing imbalances in blood sugar levels. In patients with diabetes, the hormone insulin does not give cells proper instructions to store or release sugar. This affects the entire body, as blood sugar levels must be very constant for our bodies to function properly. Osteoporosis in older women (and to a lesser extent in older men) is another well-known hormonal disorder, sometimes caused by the reduction in the concentration of sex hormones after menopause. Dwarfism, very short stature in some men and women, is caused by a lack of growth hormone. These are only a few, perhaps the best-known examples, and yet there are hundreds of different hormones in our bodies, each with its specific function that helps regulate the correct functioning of our bodies.

When we think of love, we often think about sex hormones. Sex hormones are, as their name suggests, essential for sexual reproduction. However, we do not know whether they are involved in love at all. They are crucial for sexual desire and lust. They are responsible for libido, but they most likely play only a minor role in feelings of attachment and love. Nevertheless, they are very important, as reproduction is, from an evolutionary perspective, the most essential function of our bodies. In a very prosaic way, we can say that from a selfish gene hypothesis, the only essential parts of our body are reproductive organs.

Sex organs will produce offspring and ensure the continuation of our species and the preservation of our genes. All other organs are only a necessary collaborators to enable and optimize the functioning of the reproductive system. Yet, paradoxically, looking from another perspective, reproductive organs are one of very few dispensable organs in our bodies. Nobody can survive without the heart, liver, kidneys, or most other organs. But as we have seen from the example of castrati, people could live reasonably normal, healthy, and even successful lives even in the absence of the reproductive organs.

Chemically, sex hormones are similar to cholesterol. Cells that produce sex hormones convert cholesterol into a group of male sex hormones, called androgens,[1] or two groups of female sex hormones, called estrogens[2] and progestins. The main representative of androgens is the well-known testosterone. The primary representative of estrogens is estradiol, and the main progestin is progesterone.

Chemically, all three hormones are closely related. Progesterone is an intermediate step between cholesterol and testosterone, and only one enzymatic reaction converts testosterone into estradiol. All three hormones have very different functions.

[1] Group of male sex hormones that comprises testosterone, dihydrotestosterone, androstenedione, dehydroepiandrosterone, and some others.

[2] Estrogens are a group of female sex hormones comprising estradiol, estriol, and estrone. Estrogen as a single hormone does not exist, even though the singular name estrogen is often used incorrectly in lay literature.

This specificity of their role is achieved by activating specific receptors in their target cells.

Testosterone is the primary male sex hormone. It is solely responsible for the development of male reproductive organs, the penis, the scrotum, and some internal glands during the fetus's development inside the womb. Estrogens, on the other hand, are not active in the female fetus during development, and the female genitals develop along the female developmental path by default. In the complete absence of hormone exposure and even in the presence of the male-specific Y chromosome, but without testosterone, the fetus develops normal female external genitalia and is born as a baby girl. If gonads do not develop and do not produce sexual hormones during development, the baby will be born with female genitalia regardless of its genetic make-up. Women's external organs, therefore, follow a normal developmental pathway in humans and other mammals, and men genitalia are deviation from this default pathway, caused by testosterone.

During puberty, ovaries become active. They begin to secrete large amounts of estrogens. Similarly, testes begin to secrete significant amounts of testosterone after the inactive phase during childhood. These two hormones then trigger the final transformation of the male or female child into a sexually mature person. Testosterone instigates the deepening of the voice and growth of facial hair in males. Estradiol activates the development of breasts in females and triggers reproductive cycling, starting with the first menstruation. Interestingly, the growth of pubic hair is caused by testosterone in both men and women. If, for some reason, women have excess testosterone in their bodies, this testosterone can cause the growth of facial hair. With surplus testosterone, women could also develop baldness, which is usually a problem only in males.

After puberty, testes secrete testosterone more or less continuously (although with daily fluctuations). It has the function to maintain male features and support the production of sperm in the testes. In women, however, hormones fluctuate during the menstrual cycles, and estradiol and progesterone play an essential but different roles during various parts of the cycle. After menstruation, new eggs begin to mature in the ovaries. During egg maturation, estrogens secreted from the cells surrounding the developing egg prepare the rest of the female body for eventual conception.

During the first part of the menstrual cycle, the uterine wall thickens under the influence of estrogens. It prepares to receive and nourish a tiny embryo. Immediately after the egg is released from the ovary, cells surrounding the maturing egg change and start to produce progesterone instead of estradiol. Progesterone plays a crucial role in protecting the very early embryo. While estradiol stimulates the uterine wall to thicken during the first phase of the menstrual cycle, progesterone stimulates the cells in this wall to secrete a thick fluid, sometimes called womb milk. This fluid serves as food for the newly formed embryo for several days before blood vessels from a baby and mother form the placenta. Additionally, progesterone has crucial roles in inhibiting incessant contractions of the uterus, as these contractions could otherwise expel the tiny embryo out of the womb.

Sex hormones also have other roles in our bodies, many of which we still do not fully understand. Both estrogens and androgens influence the metabolism. These hormones increase the general activity of our cells. An increase in body weight with age, especially in women after menopause, is at least partially related to the reduced levels of androgens and estrogens in our bodies. The reduction of both hormones makes our cells lazier, and they consume less energy. Saved energy is then stored in the fat tissue, which consequently enlarges. Estrogens influence our hearts, as women suffer from heart attacks less often than men, particularly before menopause.

After menopause, the risk for heart attacks increases in women. However, the overall risk does remain lower in women than in men. Estrogens also influence wound healing, the function of our immune system, and also the brain. Although we do not fully understand how estrogens and androgens affect the brain, we know that many brain diseases are different either in the incidence or how they appear between men and women. Autism is four times more common in boys than in girls. A major depressive disorder affects two to three times more women than men, and there are similar differences in many other diseases of the central nervous system. In most cases, these differences are undoubtedly the consequence of the difference in the presence of sex hormones in our bodies. However, we do not yet know and understand how sex hormones contribute to differences in brain diseases.

Very similar to sex hormones are hormones produced in the adrenal glands that regulate how we cope with stress. These hormones, whose primary representative in humans is cortisol, are also produced from cholesterol, but through different enzymatic pathways than sex hormones. The adrenal gland, a small endocrine gland located on the top of the kidneys, has two main roles. One is hormonal regulation of water levels in our body by the hormone aldosterone. Aldosterone, together with some other hormones, ensures that we drink enough fluids and do not die from dehydration. Contrary to the widespread opinion that has been growing in recent years, our bodies have excellent control mechanisms that tell us when we have to drink. We do not have to force ourselves to drink a ridiculous amount of fluids, as is sometimes recommended in magazines or on social media. Our bodies are well versed in regulating our thirst and need for liquids. These mechanisms have evolved over millions of years and are fully functional in humans and animals.

Another function of the adrenal gland is to control our response to stress, a reaction in any situation that potentially threatens us. Adrenal glands also produce small amounts of male sex hormones androgens. However, this function is most important during the development of the baby still in the womb. It is less important in adult males, as their testicles produce vast amounts of testosterone. In contrast, in women, androgens from the adrenal glands are responsible for the growth of pubic hair.

Stress is a natural and very beneficial reaction of our bodies, which enables us to survive threatening situations. Immediately after we perceive a threat, our adrenal glands release large amounts of the hormone adrenaline into our blood. This adrenaline prepares us to fight the danger or to flee from the threat. However, adrenaline is a very unstable molecule and degrades very soon after it is released into the blood. So shortly after the release of adrenaline, caused by the stress, the adrenal gland also starts to release cortisol. Cortisol has similar but more widespread and longer-lasting effects than better-known adrenaline. Adrenaline and cortisol shut down all the functions performed by our bodies that are unnecessary for immediate survival in dangerous, potentially life-threatening situations.

Under stress, the function of the reproductive system is blocked because we do not have time to reproduce while fleeing from the danger. It would be impractical if we would become hungry, or have to go to the toilet, while fighting an enemy, so our digestion slows down. The kidneys produce less urine, as it would be impractical to go to pee while fleeing from the predator. On the other hand, stress hormones stimulate the systems that can help us to survive: breathing becomes faster and deeper, so our bodies get more oxygen. The heart beats faster, so that oxygen and blood sugar reach the muscles quicker and in larger quantities. Muscles act stronger, and with more stamina, so we can run away from the danger. Stress hormones also act on the brain to sharpen our thinking so we can concentrate on the threat. They stimulate our memory, so we remember the danger for the next time.

Stress is any real or potential danger, and stress response has evolved to ensure the survival of the individual. If a lion chases the zebra, the zebra's stress response helps it run longer and faster, thus increasing her chances of survival. If we find ourselves on the edge of a cliff, the stress response helps us focus on the danger and guides us to get away from the cliff unharmed. However, the potential threats in human societies were not only predators such as bears, wolves, or lions, or natural hazards such as cliffs, swift streams, or high tides but also other human beings not belonging to our immediate group. Therefore, meeting strangers and especially getting close to them are always stressful, even if we like that stranger. Perhaps counterintuitively, while sex hormones are not involved in love, stress hormones are. One of the hallmarks of early romance is an increase in stress hormones in our bodies. We are alert and stressed, partially because we want to do something unnatural: let a stranger enter our intimate world.

On the other hand, when we fall in love, we are also awfully worried whether this other person will feel the same for us. We are concerned whether they will fall in love, whether we will meet their expectations. So many questions without proper answers cause stress. Therefore, the first weeks of love are beautiful but often very stressful, with very active adrenal glands secreting large amounts of stress hormones. However, once we have this first stressful phase of being madly in love behind us, living together in a happy relationship reduces the stress. People who live in a fulfilling, long-lasting partnerships have lower levels of stress hormones in their blood than single men or women. Serene, lasting love is thus beneficial for our bodies.

Another hormone closely related to the functioning of the reproductive system, but probably not to love, is melatonin. Melatonin is produced in a small gland located inside the brain, called the pineal gland or epiphysis. The pineal gland responds to the signals from the part of the brain that measures time in our body, a small group of cells called the suprachiasmatic nucleus inside the hypothalamus. This group of neurons has its own internal clock which consists of sophisticated molecular pathways that involve the activation of certain genes, the production of specific proteins, and the breakdown of these proteins. By precisely timing gene activation, protein production, and protein degradation, the neurons in the suprachiasmatic nucleus know what time of the day it is. Even if we take these neurons from the brain, place them in a tissue culture dish, and grow them inside the dark incubator in the laboratory, without any access to the outside light, these cells still measure time and know what time of the day it is.

Inside the body, these neurons send signals to the pineal gland, which responds to the messages with the synthesis and release of the hormone melatonin. When the daylight wanes and darkness begins to surround us, the pineal gland is switched on by the suprachiasmatic nucleus and begins to secrete melatonin. An increased concentration of melatonin tells the brain and other organs that it is dark and we should go to sleep. In the morning, with daylight, the amount of melatonin starts to decrease, and we feel awake again. Some other parts of the brain are even more directly involved in regulating sleep and waking cycles and are precise switches that send us to sleep. But these are only connected with sleep, and we will not talk about them, as they are not directly associated with either falling in love or reproduction.

Nevertheless, sleep centers could be affected by some types of love. We all know that sometimes when we are freshly in love, we may have trouble sleeping, as we incessantly think about our loved one. But this is most likely caused by stress, a frequent companion in the early love, and not directly by feelings of love.

Melatonin, however, has essential functions in the regulation of the reproductive system. Melatonin provides everyday signals to the body when it is dark and when it is light. As the length of the day and night changes throughout the year, unless we live very close to the equator, melatonin is an annual biological clock perceiving changes in the seasons. In the summer, when nights are short, melatonin is secreted for a shorter period of time. During the long winter nights, melatonin is secreted for a longer time.

The amount of melatonin, especially how long it is present in the body, tells the whole body what season it is. Knowing when it is summer and when is winter is not trivial. This comprehension is essential for many animals. For the offspring's survival, it is essential that babies are born in the season when the chances for survival are the highest. Such a season is usually spring when the weather gets warmer and food becomes more abundant. Thus, most of the wild animals time their pregnancy so they have offspring in the spring, and they know how to time their pregnancy with the help of melatonin.

An increasing amount of melatonin tells male and female deer that the days are getting shorter. Shorter days mean the autumn is approaching, and it is the time for the rut so babies will be born in the spring.

On the other hand, in hamsters with much shorter pregnancies, decreasing secretion of melatonin at the end of the winter tells these animals that it is time to mate and have offspring in the spring. In horses, the pregnancy lasts a little less than a year. To horses, falling amount of melatonin in the spring due to longer days tells their reproductive system it is time to become active and reproduce, even though the babies will be born only after almost a year.

So, melatonin does not act in the same way in all animals but in different ways to achieve the same result. In deer, more melatonin stimulates the reproductive organs to become active and ready for mating. In horses, hamsters, and other rodents, less melatonin in the spring due to the shortening of the nights has the same role, instructing reproductive organs to become active.

We are not "seasonal animals," and we reproduce all year round. Still, we often speak that spring is the season of love. When nature awakens after the winter, we feel rejuvenated and ready to fall in love. Warm sun, colorful flowers, the buzzing of bees, and pretty scents all stimulate our senses, and we are prepared for love. But maybe it is not only the spring outside that makes us feel different. Hormones, not only melatonin but also testosterone, estradiol, and dopamine, fluctuate in our bodies throughout the year. Our bodies react to stress differently during the dark and cold winters than in colorful and warm springs. All these could contribute that we are more prone to fall in love in the spring, although this is not scientifically proven. It could be only because we are bombarded with the messages from various media that spring is the season for love, and we do not feel differently because of our hormones.

Of course, humans also have melatonin which tells us when it is day and night. However, with electric lights and other modern devices, we can cheat this system. Whether melatonin has a role also in the functioning of our reproductive system is not yet known. As it has essential roles in most animals, it is likely that melatonin also has some role in humans. Therefore, melatonin, sometimes advertised as a miraculous drug, should be taken with great care, especially in children. Melatonin is sometimes used in children during puberty to help them with insomnia. Sometimes, people use melatonin to combat jet lag, and some people use it even as an anti-aging drug, without any evidence that melatonin could slow down the aging. This might not be the wisest of things to do. We still do not know enough about what melatonin does in human bodies. Therefore, we should be careful when prescribing such hormonal drugs, especially to the most vulnerable members of our society, the children.

Two other hormones, prolactin and oxytocin, are thought to be profoundly connected with attachment and love. Both are secreted from the pituitary gland and have their primary functions during pregnancy and breastfeeding. Prolactin stimulates the milk production, while oxytocin stimulates uterine contractions during labor. Later, oxytocin also constricts the milk sacs in the breasts to release milk for a baby.

These two hormones have essential roles also in the brain where they can trigger the feelings of attachment between mother and her child, which is probably the evolutionary origin of romantic love. We will, therefore, talk much more about these two hormones later. Vasopressin, a hormone that is very similar to oxytocin and is also released from the pituitary, has a primary role in regulating fluid balance in our body. It constricts the blood vessels in the kidney to prevent the excretion of fluids when we are dehydrated. However, vasopressin also acts in the brain. In monogamous animals, it triggers feelings of attachment together with oxytocin. In humans, it has a role in sexual arousal. Although its role in the feelings of romantic love is not clear, it is most likely involved in the attachment and bonding between lovers.

As seen with oxytocin and vasopressin, the same molecules can have several different roles in the body. Often, some molecules act as hormones and are carried by the blood throughout the body. In the brain, the same molecules are mediators in communications between the nerve cells. Neurons communicate with each other by releasing neurotransmitters, substances discharged from one nerve cell that tell another nerve cell what to do. Neurotransmitters act through similar receptors that are also present on other cells in our bodies, where they are essential for the recognition of hormones. The utilization of the same molecules and their receptors is another similarity between the nervous and hormonal systems.

Many neurotransmitters are similar to hormones, and many are even the same substances as hormones. Nature is thrifty and uses the same substances in different ways, whenever possible. Sometimes, as with oxytocin and vasopressin, it is difficult to distinguish between a hormone and a neurotransmitter. A general rule is that hormones are substances that are released into the blood and are transported by blood around the body. In contrast, neurotransmitters are released by nerve cells and act only on neighboring neurons.

So, when we talk about oxytocin, we talk about hormone and neurotransmitter at the same time. When oxytocin triggers the contraction of the womb, it is considered a hormone. If the same substance acts in the brain and triggers the feelings of attachment, it is appropriately called a neurotransmitter. Therefore, when we talk about oxytocin, it is sometimes difficult to distinguish its role as a hormone and as a neurotransmitter. The word hormone is often used even when we think about oxytocin as a neurotransmitter.

Hormones and neurons regulate the function of our entire body, including how we feel and what we think. There are probably more than 100 hormones in our bodies, and different hormones control different parts of our bodies. Hormones have profound roles in the functioning of the organisms. They are responsible for how we feel, how we react, how sleepy we are, and how productive we are. Hormonal imbalances are behind depression, anxiety, diabetes, and many other diseases. Hormones are responsible for acne and many cases of infertility. Hormones determine how tall we grow during puberty, and some hormones even regulate the growth of certain cancers.

Hormones are also deeply involved in both love and reproduction. Sex hormones, stress hormones, and melatonin all play important roles in regulating the reproductive system. Without sex hormones, our reproductive system does not function. Stress hormones temporarily shut down the reproductive system so that we can flee from danger without distractions. Melatonin tells animals when it is the best time to mate so that the offspring will have the highest chances of survival. These hormones are certainly involved in physiological responses leading to reproduction, but their role in love is less clear.

The lust or desire for sexual reproduction, which is regulated by sexual hormones, is part of love. Being madly in love is definitely stressful, and the length of the day and night influence us profoundly. Perhaps it is not surprising that we celebrate Valentine's day in February when days grow longer and spring is approaching. We do not detect the end of the winter only by seeing the green grass, smelling fragrant flowers, listening to the buzz of bees, and hearing melodious birds' songs. Our bodies detect spring also by falling levels of melatonin, and this perhaps makes us more prone to fall in love. But feelings of romantic love and attachment do not come from testosterone or estradiol or melatonin. Other hormones and neurotransmitters such as oxytocin, vasopressin, and dopamine, produced and acting inside our brains, have more important roles in determining with whom and when we fall in love. These neurotransmitters affect our feelings and emotions. They seem to be behind our feelings of love, closeness, and attachment to our intimate partners.

While testosterone triggers the desire for sex, oxytocin, vasopressin, and dopamine will ultimately usher us into loving our partner. They will convince us to stay with our beloved person and make us long for the happiness we experienced when we were with our love. After all, as George Sand said: "There is only one happiness in this life, to love and be loved."

# Chapter 7
# Beware, an Elephant in the Musth

*Today is World Elephant Day and I am thinking of the magical times when I have been privileged to watch wild elephants in different places in Africa. How wonderful it was to see them feeding, using their trunks like hands to pick leaves overhead, or tear up lush green grass in great bunches, then push the food into their mouths. And special times when I watched them bathing in a river, sucking up trunk full of water and squirting it over their backs, sometimes submerging with just the tips of their trunks showing like periscopes above the water. And what fun the young elephants have playing in the water, squirting each other and messing about like kids in a swimming pool.*

Jane Goodall

A few years ago, on a family safari in the Kruger National Park in South Africa, I startled a young elephant bull while driving around at dawn. The bull elephant had his breakfast, peacefully munching leaves and branches. When we came around the corner, although I was driving slowly, our car startled him, and he became upset.



G. Majdic, *Soul Mate Biology*, Fascinating Life Sciences,
https://doi.org/10.1007/978-3-030-67212-6_7

Suddenly he stood in the middle of the road, trumpeting loudly and swinging his head from side to side. It was not a pleasant sight. The elephant was much larger and heavier than our car, and angry elephants occasionally attack cars. Fortunately, I managed to change into the reverse gear, and the elephant did not chase us once we moved away. The situation was quite tense, because young elephant bulls could sometimes be dangerous.

After everything had calmed down and we drove on, I thought that his reaction was actually quite familiar to me. In the back seat of our car was my teenage son. Although we go on together very well, my teenage son's behavior is occasionally just like that elephant's behavior. An outburst of anger coming from nowhere, screaming, shouting for a short time, and seemingly for no apparent reason until things cool down. We should never try to see our behavior in the behavior of animals. But in this particular moment, I could not help but smile at the thought of how adolescent boys are not similar only across cultures but perhaps also across species.

Elephants are perceived by most of us as gentle giants, strolling slowly and harmlessly around African savannah, taking care of the cuddly young babies, gathering in herds, and irradiating gentleness and tranquility into their surroundings. However, male elephants can be very opposite of tranquility and peacefulness. They can become very aggressive and dangerous at certain periods of life. When male elephants reach around 15 years of age, they usually separate from their original herds and socialize mostly with other males. They like to keep the company of older males that can serve as their tutors, in good and evil. Several years later, young bulls enter their first musth, a unique state during which male elephants transform from placid giants to highly dangerous behemoths.

Male elephants in musth became very asocial, easily disturbed, and aggressive. They are upset by loud noises or sudden movements and are ready to attack anything that moves or makes a sound. In the wild, they would attack other elephants, not just bulls but also peaceful cows. They might attack cars if vehicles get too close to them. On rare occasions, young elephant bulls in musth even attack human settlements. Elephants in captivity attack other elephants as well as their keepers when in musth. They are extremely dangerous during this period, which in adult elephants often lasts several months. This unruly behavior is causing a big problem for zoos, as bulls in musth must be separated in individual enclosures and cannot be approached by the keepers during the musth. That is the main reason why most zoos only keep elephant cows.

Blood levels of the male sex hormone testosterone can be up to 40–60 times higher in bulls in musth than in bulls that are not in the musth. This is most likely causing both behavioral and physiological changes. However, it is not clear whether the testosterone is triggering or causing the musth or high levels of this hormone are just one of the musth's consequences, triggered by something else.

Elephants in the musth are easily recognized. They have swollen glands on the temples, and these glands secrete dark oily liquid that drips down the head of the elephants. Bulls constantly dribble urine and have permanently wet hind legs. They emit a strong smell of urine, and the foreskin of their penis becomes white-greenish color. They are very bad-tempered, as swollen glands on their heads presumably cause an intense pain that might be similar to a stabbing toothache in humans. We all know how irritable we can be with a painful tooth, so we can have some understanding for elephant bulls.

What is the purpose of musth is still a mystery. It is not, as many people think, similar to the rut in deer, which takes place during the mating season of the stags. Elephant cows are not seasonal breeders. They can enter estrus and mate with bulls at any time of the year, and they mate with bulls in musth and with bulls that are not in the musth. Musth might be important in the fighting for male dominance or establishing a territory. Perhaps musth is giving a chance to mate to smaller and weaker bulls and thus increases genetic diversity. Although bulls not in the musth could also mate if they find a willing female, cows usually mate with bulls in the musth, as these males are the most lustful due to high levels of testosterone. Males in musth commonly win all fights between bulls, even if the rival not in the musth is more massive and might appear stronger.

The musth is not very comfortable for bulls, but it has important roles. It signals to the cows which males are strong and healthy and ready to mate, as weak and sick bulls do not enter the musth. Similarly, antlers in deer, colorful plumage in birds, and facial hair and deep voice in humans all serve as signals to tell the females of the appropriate species that males are mature and ready to mate. Some of these signals, such as musth in elephants or antlers in deer, are only present during certain periods of the year. However, they occur repeatedly, either regularly like the antlers in deer or irregularly like the musth in elephants. Some signals like the beard in humans or mane in lions are present permanently. These signals, whether permanent or recurring, share several common features. One is that they are all influenced by the male sex hormone testosterone. Another one is that they all occur for the first time during puberty, a period of life when sexually immature animals or human teenagers become sexually mature and ready to reproduce.

In our societies, the notions of love are present with kids from early on through children's stories, nursery rhymes, and cartoons. Young children may already talk about being in love in kindergarten. But young children are neither ready to reproduce nor to love in the same sense that we talk about love in adults. Only during puberty the organism matures and becomes prepared to mate and produce offspring.

The time before different animals and humans reach sexual maturity is very different in length. In general, the larger the species, the longer time it takes to reach sexual maturity. Mice are sexually mature already at 2 months of age, and dogs usually around 8–10 months of age. Cows reach sexual maturity between 12 and 18 months of age. Humans and elephants are sexually mature much later, sometime between 12 and 15 years of age. This lag period is essential for the development of the organism. The body, especially female, has to grow and develop to be able to support the childbearing. Animal males have to grow to be strong enough to fight other males to win the love of the females.

During puberty, reproductive organs that are quiescent throughout the childhood years become active. All differences between men and women become apparent, mostly caused by the action of sex hormones secreted by gonads, testosterone in boys and estradiol in girls. The function of the reproductive system is regulated at the highest level by a hormone called gonadotropin-releasing hormone, or shortly GnRH. A few hundred specialized brain cells in the hypothalamus produce GnRH.

When released, GnRH stimulates the pituitary gland to secrete two other hormones, the luteinizing hormone (LH) and the follicle-stimulating hormone (FSH). These two hormones stimulate the gonads, testicles in men and ovaries in women, to produce mature germ cells, sperm and eggs. Gonads also produce sex hormones, testosterone in males and estradiol in females. Testosterone and estradiol are the final activators that regulate the development of all differences between males and females during puberty. Testosterone causes a more muscular body and triggers the growth of facial hair, an enlargement of the larynx, the appearance of the Adam's apple, and a deeper voice. Estradiol controls the development of the breasts, thicker and smoother skin, and different distribution of body fat and muscles. An increase in estradiol during puberty also triggers the commencement of menstrual cycles and the first period. All this starts to happen only during adolescence, as before puberty, neurons that produce GnRH are inactive.

What triggers puberty is still not known. One of the essential steps toward puberty is that there are enough energy sources stored in the female body. These energy stores are needed for the sustention of pregnancy, during which the energy requirements of a developing child are very high. Fat tissue in our bodies secretes a hormone called leptin, which signals to the brain how full our energy stores are. The more fat we have stored, the more leptin we have in our blood. During puberty, parts of the brain that control sexual maturation also monitor leptin levels. If leptin levels are not sufficiently high, the brain does not activate the process of sexual maturation, even if the age is right to reach sexual maturity.

Although we know that a sufficient amount of energy stores in the form of fat is needed for a human body to enter puberty, this is not a trigger. It is only a permissive step, a threshold that must be reached before puberty is allowed to occur.

A molecule with the poetic name kisspeptin is crucial for the onset of puberty, as human patients with the defect in this gene do not enter puberty. Its poetic name would be more than appropriate for a gene that triggers puberty, causes sexual maturation, and eventually leads to love. But kisspeptin was not named after a passionate kiss. It was initially discovered in Hershey, Pennsylvania, home of the famous chocolate factory where Hershey's chocolate kisses are made, and hence the name. Kisspeptin is present in the hypothalamus, and neurons producing kisspeptin are in direct contact with the neurons producing GnRH. But we still don't know what switches on the kisspeptin to become active during puberty. Therefore, we still do not understand what triggers puberty.

During puberty, important and well-known transitions occur. In humans, besides the development of secondary sexual characteristics, a growth spurt is one of the puberty hallmarks. All parents know how their children seem to grow almost overnight, especially boys, changing from a child to a tall young man. But of course, this is not the most important thing about puberty. As every parent of a teenager knows or remembers, puberty also brings major changes in the teenagers' temperament and behavior. The reason for this is the rapid brain development and brain remodulation that occurs during this period of life. Mostly under the influence of sex hormones estradiol and testosterone, which are now actively secreted from the gonads, the brain undergoes significant changes.

One of the most noticeable changes in the brain is the increase in the myelination of the nerve fibers. Nerve cells communicate with each other through their protrusions, called axons and dendrites. Axons and dendrites are long and thin fibers, extensions of nerve cells, used to transmit electrical signals from one nerve cell to another. The specialized coating called the myelin sheath envelopes all nerve cells. This helps electrical signals to travel faster and they are not lost or reduced on the way from one cell to another. The myelin sheath has pretty much the same function as the plastic coating of electrical wires in our homes and electrical appliances. The myelin sheath ensures faster and more accurate communication between different parts of the brain.

During puberty, extensive myelination of nerve fibers occurs, which improves the functioning of the entire brain. Yet this does not happen in the whole brain simultaneously but gradually. Myelination during puberty begins in the parts of the brain at the back of the skull and progresses slowly toward the front. The area that lies just behind our foreheads, the prefrontal cortex, goes through this vital process that allows fine-tuning of our brains the last. The maturation of the prefrontal cortex continues long after the beginning of puberty. This area may not be fully mature until 24 years of age.

Delayed maturation has significant behavioral consequences. The prefrontal cortex is an area of the brain that executes numerous higher cognitive functions. This part of the brain is the primary site where we focus attention, organize thoughts, and plan how to solve problems. The prefrontal cortex also inhibits inappropriate behaviors, analyzes complex situations, and foresees the consequences of our actions. It adjusts the behavior according to the situation, modulates intense emotions, controls (keep in check) impulsive behavior, and balances long- and short-term goals and rewards. This list must sound familiar to most of the teenager's parents, as many of such practices, or better lack of, are at the base of the troubles parents often have with teenagers.

On the other hand, areas of the brain associated with emotional responses develop more rapidly. They are very active during puberty and early adolescence. One of these areas is the amygdala, a small almond-shaped part of the brain. The name amygdala means almond in Greek. So, do not be surprised if you travel to Greece and find that amygdala is sold in the shops. They do not sell parts of the brain but regular, tasty almonds.

The amygdala is located below the cerebral cortex and is one of the primary brain regions responsible for processing emotions such as excitement and fear. When we see something that we have learned to fear, the amygdala generates feelings of fear and anxiety. When we are elated with positive emotions, the amygdala is active. As we are conscious beings, we usually do not act instinctively but keep our positive or negative emotions under control with the help of the prefrontal cortex. Any insult that would cause an emotional upset in the amygdala is, therefore, in adults, usually held in check by the prefrontal cortex. The prefrontal cortex makes sure that our emotional response is not exaggerated and inappropriate.

In the teenagers' brains, however, the connections between the prefrontal cortex and the amygdala are not yet finely tuned, which causes discordance in the adolescent brain. Any emotional trigger produces a strong activation of the amygdala in the teenager's brain. As the prefrontal cortex and especially the synaptic connections between the prefrontal cortex and the amygdala are not yet fully matured, these emotions are expressed as intense emotional outbursts, with which all parents of teenagers are well acquainted.

There are similar problems with other parts of the intuitive brain, like the reward system. A reward system is a circuit in our brain that gives us a feeling of pleasure and self-content when we do something that is essential for our survival or survival of the species. When we eat high-calorie food (food with lots of sugar or fat), we provide our body with energy that helps us survive. When we have sex, potentially producing an offspring and thus ensuring our genes will survive, we are doing something essential for the survival of the species. In both instances, the reward system is activated by the neurotransmitter dopamine and gives us a pleasant feeling of fulfillment.

In adults, the prefrontal cortex also keeps the reward system in check, and we can be satisfied with little pleasures like a yummy piece of chocolate cake. In teenagers' brains, these connections are not yet fine-tuned, and teenagers often seek greater pleasures that can be provided by different drugs and risky behavior. This proneness to risk-taking behavior is the cause of a higher vulnerability of teenagers to drug abuse and undertaking various risky behaviors.

We all know that teenagers are more likely to drive fast and cause traffic accidents. They engage in vandalism and truancy. They have unprotected sex and indulge in binge drinking. Interestingly, undertaking risky behavior is not unique to humans. Young male elephants often venture into the fields, damaging crops and exposing themselves to danger. Young male gazelles often approach and tease cheetahs and lions. Young male sea otters swim to the open sea, away from the protection of the kelp, and expose themselves to the white sharks. Adolescent male gorillas try to pick fights with stronger and larger opponents. Immature male antelopes engage in bouts with sharp horns and often damage each other.

In humans, risk-taking behavior is not exclusively the domain of boys; but it is more often undertaken by men than by women. The same applies to animals, so it seems there is something special about the male teenage brain. At least partially, it is thought to be due to evolutionary pressure on males to fight for the living space and females and is triggered by elevated testosterone levels during puberty. Adult male animals have to establish their territory and their place in the community. This is often achieved through the fighting between males. Although many adult animal males do not engage in actual fights but settle their scores through threatening postures, young males have to learn how to fight for their place in the community. Reckless pubertal behavior is part of the learning process and maturation, and such behavior in our kids is a remnant of old evolutionary pressure on males.

In human and animal adolescents, peer pressure has a crucial role in risky undertakings. Teenagers' brains are particularly susceptible to judgment and praise from their peers. Even in straightforward tasks, adolescents undertake risky behavior more often when in a group of friends. Young elephants usually raid the crops in a company. A small group of adolescent male gazelles often chase lions. Human teenagers more frequently run across the street through the red light if they are in the company of their peers than if they are alone. They usually drive dangerously fast when they try to impress someone. The same is true for drug-taking, binge drinking, and other dangerous activities. It seems that the two processes are going on simultaneously. One is a lack of judgment and deficient experiences due to immature brain connections. Another is an urge to impress the peers, win the females' hearts, and defeat potential contestants for our place in the community.

Another part of the brain that undergoes significant changes during puberty is the hippocampus. This area is connected with the memory, in particular with the consolidation of memories. Consolidation of memories is a process during which the brain decides which memories will be stored in our brain for a very long time and which recent memories will be erased from our brains. The hippocampus is one of the brain areas where new cells are born throughout life, but the occurrence of new cells is especially prominent at puberty. Many new cells are born in the hippocampus during puberty. This increases the brain's capacity to learn new things and store new memories. Such ability is of paramount importance during this rapidly changing period of our lives when we learn and gain experiences for the rest of our lives.

Many cells are born every day in the hippocampus, and many also die. Some deaths are normal, because more cells are born than are needed. However, various external factors can influence the survival of the cells. Stress promotes the death of hippocampal cells and reduces the occurrence of new cells, hence the harmful effects of stress on learning and memory. On the other hand, learning new things supports the survival of new cells. Studies in animals and humans have convincingly shown that more cells survive in the hippocampus if we are learning something new. But it is not only the brain activity that promotes the survival of cells. Physical activity has a similar effect. Daily exercise with an increased heartbeat and faster breathing directly promotes the survival of the cells in the hippocampus.

This is true throughout our lives; therefore, we must be physically and mentally active as we age. Both physical and mental activities promote the survival of cells in our brains, particularly in the hippocampus. During adolescence, learning and physical activity have even more critical roles. While stress during puberty or earlier in childhood permanently affects the number of cells in the hippocampus and their interconnectivity, exercise and learning prime the hippocampus to function better in adult life. Teenagers who are both physically and mentally active during puberty often have better cognitive abilities throughout their lives. Exercising and learning new things during adolescence might even overcome some effects of earlier stress on the hippocampus and improve its function for the rest of life.

Puberty is a stormy and turbulent period during which our bodies mature. The gonads become capable of producing sperm and eggs that will ensure progeny. Sex hormones secreted by gonads cause changes in our bodies to transform us into two very distinct genders. The brain undergoes essential changes that will usher us into adult life. During puberty, all brain parts involved in emotional responses, including feelings of love and attachment, undergo substantial changes. They reshape, grow, and mature to prepare the brain for its functions in adult life. This includes more permanent and stable love instead of stormy, unstable, lustful love that follows us through the teenage years. We should keep in mind that teenagers are not moody and emotional and sleepy and unresponsive or take inappropriate risks just to upset the parents. Most of these, sometimes annoying, behaviors have a biological basis in rapidly developing and reshaping the brain.

We also firstly fall in love during puberty, but how and why this happens is still largely a mystery. Increased levels of sex hormones surely promote lust and sexual desire. But again, lust is not love, and testosterone and estradiol do not have essential roles in romantic love. There are some suggestions that levels of oxytocin and vasopressin in our brains change during puberty. However, we know that both are present in the brain from very early life, as oxytocin is also essential for the feelings of attachment in babies and children.

In puberty, we switch our attention from parents to our peers. We are social beings. Before puberty, we primarily seek comfort and attachment with our parents. In adolescence, we start to sever these ties and seek social fulfillment elsewhere. Our reward circuits are more active during puberty than in adult life, and love provides immense pleasure by activating dopamine in the brain. With the conscious prefrontal cortex idling to some extent, all this is a recipe for a perfect storm. With high lust from testosterone or estradiol, high levels of dopamine, and no prefrontal cortex to hold the reins, teenagers are falling madly in love swiftly and effortlessly. Yet, teenagers also often break up their relationships when the initial excitement subdues and seek fresh exhilaration with a new partner.

For all worrying parents: this is entirely normal behavior. We all went through these phases, as such behavior is embedded in our genes and in our brains. It is part of the natural processes of brain maturation and is needed to properly develop teenagers' minds. Eventually, when the prefrontal cortex matures several years later, it begin to control our emotions. Finally, teenagers learn how to enter and sustain more stable, fulfilling relationships.

We do not understand what changes in our brain during puberty to shift our focus of attachment from parents to lovers. There is undoubtedly some kind of switch in our brains that is flipped during puberty. But what is turning on the ability to feel and desire romantic love, what is the trigger for writing poems, what sparks us to write love letters, we do not know. For the time being, this remains one of the mysteries of our brains.

# Chapter 8
# Crocodile Brain

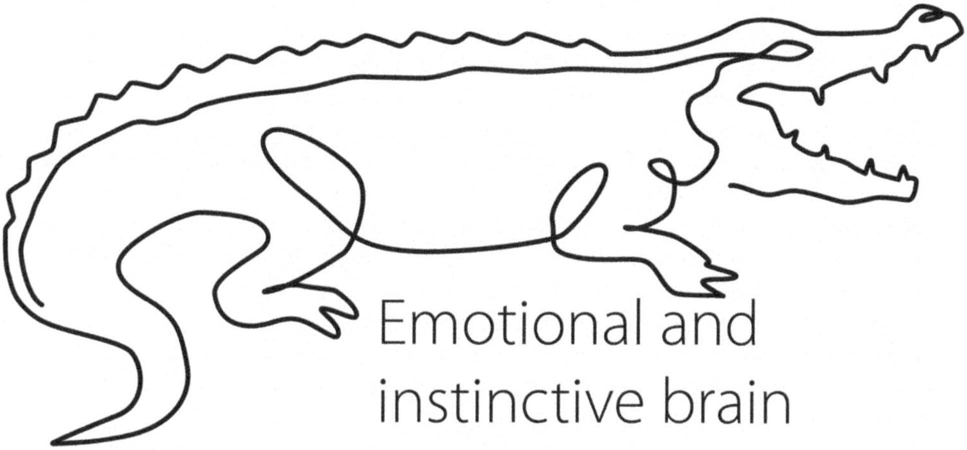


*You have brains in your head. You have feet in your shoes. You can steer yourself any direction you choose. You're on your own. And you know what you know. And YOU are the one who'll decide where to go....*

Dr. Seuss: Oh, the Places You'll Go

Although Dr. Seuss wrote many immortal and truthful quotes, the one above is not entirely correct. In most circumstances, we are in control of our brains. Or better said, our brains are in control of our brains and, consequently, of our actions. We can choose when to eat and when to go for a stroll. It is our conscious decision when we read a book or when we go out with friends. We decide when to kiss our loved one. But sometimes, we are not in control of our brain and our body. This sounds very paradoxical, as, in a sense, we are our brains. Occasionally, parts of this complicated organ take over our consciousness and do things that might surprise us. Instinctive parts of our brains send electrical and chemical signals around our bodies without our conscious mind being able to control them.

When we traveled in South America, my friends and I walked on the slopes of the Purацé volcano in Colombia. High up on the slopes, we had to cross the pasture, and suddenly, a seemingly friendly cow started to run toward us, looking less and



G. Majdic, *Soul Mate Biology*, Fascinating Life Sciences,
https://doi.org/10.1007/978-3-030-67212-6_8

less affable. We had no time to think. We just turned around and ran as fast as we could (which was not too fast since we were at high altitude). We jumped over the high fence without thinking about how we did it until we were finally able to rest with a strong barrier separating us from the cow. This escape was an instinctive reaction to the perceived danger, a natural instinct embedded in our brains and muscles. I am not afraid of cows, and I generally do not think of them as dangerous animals. I often meet them when hiking in the mountains, and I never run away or try to avoid them. But my mind perceived the running cow on Puracé as a danger and told me to run away without my brain thinking about it.

When we are scared, we jump or run away from the perceived threat without thinking about our actions. If we are afraid of snakes, just hearing a hiss might cause an involuntarily hop away. If we see something we think is disgusting, we will turn away and not want to look at it. We might even involuntarily vomit when we see something that we think is disgusting. If we burn a finger, we quickly move the whole arm away without thinking how and why. And finally, perhaps only once in our lifetime or maybe a few times in our lives, we see one particular person and are unable to avert our eyesight or stop thinking about this person. We get butterflies in our stomach when we think about this person. We feel anxious when we are away from that person, and we are desperate to see them again as soon as possible. We fall in love.

The actions described above are all instincts that are deeply rooted in our brains, in the evolutionary old parts of the central nervous system. In humans, this primitive part of the nervous system is usually under the control of our conscious mind, but sometimes, the instinctive brain takes over the steering wheel. Such instincts are necessary for our survival, as conscious analysis of the situation is sometimes too slow. Occasionally, we need a quick, autonomic response to flee from the mortal danger. Fear, fight, and flight emanate from the autonomous part of the brain that instinctively protects us from the ensuing peril and ensures the survival of the individual. Sometimes, however, the instinctive brain also takes command to ensure the survival of the species, not just individuals. It forces us to fall in love, what should ultimately lead to the generation of progeny. Many of the same brain centers that regulate the fear and stress response are also active when we fall in love.

The brain began to develop during evolution as connections between a few nerve cells way back in history. Our very early ancestors, unicellular choanoflagellates[1] that lived in the primordial oceans, already possessed many components needed to transmit electrical signals around 850 million years ago. Communication between a few nerve cells most likely occurred initially in the ancestors of today's jellyfish and sea anemones more than 500 million years ago. Over millions of years, an initial network of few cells with extremely limited function developed into a large gray lump of tissue inside our skulls. This consists of tens of billions of cells that are intimately interconnected into an enormous, intricate grid that controls not only

[1] Single-celled organisms with some cellular characteristics similar to animal cells. They might be the earliest ancestors of all animals.

basic instincts but also empowers us to think and create. But, as nature is always thrifty, it does not forgo useful things. Parts of the evolutionary old brain, which controls the basic instincts necessary for survival, are still present in our brains.

Today sometimes disputed triune brain theory suggests that our brains are divided into three primary functional regions. The oldest of these is the reptilian or crocodile brain, also called archipallium. This part regulates essential life functions such as breathing, heartbeat, food-seeking behavior, and to a certain extent, aggressive behavior necessary to protect the food and territory. Indeed, most of us would probably agree that these are the main things crocodiles do in their lives.

According to the triune theory, the second part of the brain is the so-called paleomammalian brain, or, more simply, the old mammalian brain. This part began to develop in early mammals about 250 million years ago. The old mammalian brain controls emotions and instincts such as joy, fear, fighting, feeding, and social and sexual instincts. It also has essential roles in learning, motivation, and memory. It consists of numerous groups of cells called brain nuclei, which are located under the cerebral cortex, and is commonly called the limbic system. This system is intimately involved in the feelings of love and attachment, together with some higher brain centers.

The evolutionary youngest part of the brain is the cerebral cortex, the neomammalian brain, also called neopallium. This part developed rapidly after dinosaurs were wiped off the Earth and mammals began to conquer the world. The cerebral cortex is present in all mammals, but it grew spectacularly large in primates and especially in us humans.

The cerebral cortex is the largest part of our brain. It is responsible for analyzing and processing all the information coming from inside and outside of our bodies. This part of the brain is responsible for thinking, reasoning, evaluating, foreseeing, and all other higher cognitive functions we perform. It is the site of consciousness, even though we still don't know how to clearly define what consciousness is. One of the major parts thought to be responsible for thinking and conscious decision-making is the prefrontal cortex of the brain, an area right behind our foreheads.

Although there is ample evidence that the prefrontal cortex is the main site of cognitive functions, it is difficult to establish such sharp distinctions between parts of the brain. The brain is a complete organ, and numerous brain centers control all our actions. Although we know exactly which part of the brain is responsible for extending our index or middle finger, this part of the brain almost never acts independently. It is always in close communications with various other parts of the brain, and so it is true for the whole brain and all its actions.

Likewise, not all our emotions are produced in one single brain center. They are modulated by the interaction between several parts of the brain. One exception to this are reflexes, which protect us and are regulated by straightforward nerve circuits. If we inadvertently touch a sharp needle with our finger, we automatically move the whole arm away from the danger. If the doctor hits us under the knee with a small hammer, our lower leg springs upward to protect the overstretching of tendons in our knees. Circuits consisting of only a few neurons control such reflexes without the involvement of the sophisticated brain centers. These circuits go from the skin (on the finger) or the tendon (knee) to the spinal cord. In the spinal cord, the sensation of pain is transferred to the neurons, which automatically move the part of the body away from the danger, often without involving the brain entirely. However, such reflexes are rather the exception, and interactions between several parts of the brain control most of the functions we perform.

The triune theory of brain division is most likely too simplistic. The whole cerebral cortex, including the prefrontal cortex, is closely connected with the reptilian and paleomammalian brain. Similarly, reptilian and paleomammalian brains communicate extensively with each other, making a strong division between these three parts of the brain difficult. The whole brain, including both the most primitive reptilian brain and the highly sophisticated primate cortex, control all our actions. The prefrontal cortex, the site of consciousness, is charged with the task to ensure that most of our activities are appropriate and put into context.

But sometimes the limbic system, the primitive mammalian brain, takes over and forces us to do things that may not be appropriate. Fighting with our competitor for the person we love, running away from a threat like a headless chicken, or constantly harassing the women we are in love with, even though she does not share our feelings, are all the doings of the limbic system. The limbic system sometimes overcomes the rationality that is normally created in our conscious cerebral cortex.

Three main parts of the limbic system are the hypothalamus, the amygdala, and the hippocampus. The hypothalamus is located at the base of the brain, deep inside

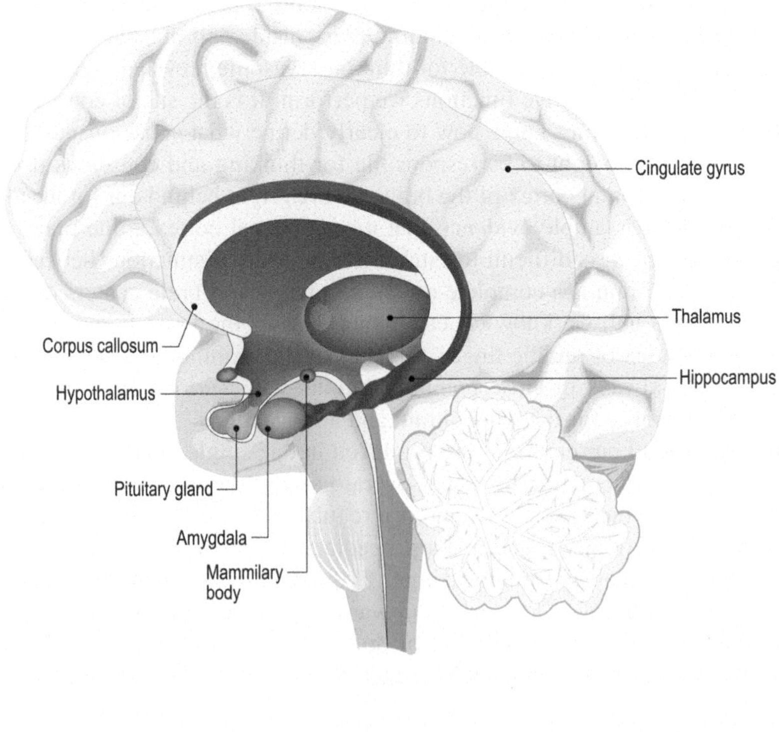

our skulls. It is considered the very center of the autonomous brain. The hypothalamus controls more or less all important autonomous functions of our bodies except essential life functions such as breathing and heartbeat. Although the hypothalamus also modulates the speed of our pulse and the depth of our breaths, it does not have direct control over whether we breathe and whether our heart beats. It can only modulate the speed and depth of these two basic life functions.

The hypothalamus is the main center for the perception of hunger or satiety by receiving hormonal signals from fat tissue and stomach. Parts of the hypothalamus trigger defensive behavior when we feel threatened. Other parts of the hypothalamus spark our sexual instincts. The hypothalamus most likely governs our sociability. It has groups of cells that control thirst and body temperature. If specific cells in the hypothalamus perceive higher than normal concentration of salt in the blood, the neurons in the hypothalamus create a sensation of thirst, and we instinctively begin to seek water. When we are outside in the cold winter, the hypothalamus stimulates muscles to shiver and produce more heat. A small group of cells in the hypothalamus, called the suprachiasmatic nucleus, measures time. These cells accurately measure time and tell the rest of the body when it is time to sleep. Parts of the hypothalamus regulate the heartbeat and breathing.

Last but not least, the hypothalamus is the gatekeeper of the entire endocrine system in our body. The hypothalamus controls the function of all endocrine, hormone-secreting glands in the body via the pituitary gland, a small gland located just beneath the hypothalamus at the base of the brain. As such, the hypothalamus determines how tall we are, when we enter puberty and start to reproduce, when we are under stress, and all the other hormonal functions.

The hypothalamus also has close connections to other parts of the brain, especially the brain cortex. Interactions between the hypothalamus and the cognitive regions of the cerebral cortex control our autonomous behavior and ensure that we behave properly in the contexts of the situations in which we are. The hypothalamus may encourage us to gobble up as much food as possible if we attend a reception with lots of tempting fare. But the brain cortex puts a brake on the hypothalamus by telling us to behave appropriately for the occasion. If we meet a beautiful woman, the hypothalamus might prompt us to start courting and even make inappropriate advances. But the brain cortex slams the break and tell us that such behavior is not appropriate. That is why complicated connections between the autonomic hypothalamus and the conscious cerebral cortex are essential for a civilized life in our societies.

The hypothalamus is the primary regulator of reproductive function. Several parts of the hypothalamus are involved in the regulation of sexual instincts. Other parts of the hypothalamus are responsible for socialization and bonding between human beings, including mother and child. Thus, the hypothalamus is profoundly involved in the feelings of love. It is most likely one of the brain's central parts that creates feelings of being madly in love when we first fall in love with someone. It also creates the feelings of togetherness and attachment that develop later during the loveable relationship.

Another vital part of the limbic system is the amygdala, an almond-shaped part of the brain, located under the cortex on the brain's left and right side. The amygdala controls our emotions, and all the intense feelings we perceive come from this part of the brain. When we see something we are afraid of, signals from the eyes travel to the amygdala, and the amygdala triggers the fear response.

If we walk on a mountain path and step near a poisonous snake, our eyes or ears (if we only hear the hiss) send the signal to the amygdala. The amygdala immediately recognizes the threat, and we instinctively jump away. The amygdala has direct communications with parts of the brain that control our muscles and trigger instinctive movements without us being aware of these moves. However, the amygdala could also temporarily stop parts of the brain that control the muscles. We all know that sometimes when we are terrified, we freeze and cannot move. The amygdala triggers this freezing, which is an ancient response to threats. Although in the context of human behavior, it might seem counterintuitive to stop and not run away from the danger; freezing is a good tactic for prey to avoid being eaten by the predators.

Many predators rely on the detection of movement when hunting, and freezing of prey on the spot confuse them. If we find ourselves face to face with a lion or bear in the wild, we should not sprint away, as this quick movement instinctively turns the predator instinct on in the predatory animal. Similarly, if an owl chases the mouse, mouse freezing ceases giving various clues, such as the movement of tall grass. This lack of movements makes it more difficult for the owl, a predator, to find its prey, a mouse. Freezing behavior is so effective and useful that mouse under attack secretes pheromones, which automatically trigger freezing behavior in all nearby mice, thus protecting them from a predator.

Amygdala also has a vital role in our memory system. Although one of the main brain areas involved in memory storage is another part of the limbic system, the hippocampus, the amygdala also stores some memories. These are memories of dangerous, stressful events that could be threatening our lives. When our ancestors ran away from the lion somewhere in the African savannah, it was useful to remember that the lion is dangerous and should not be approached again. The amygdala stores these kinds of memories, and the activated amygdala, when we are scared, has a better ability to save the memories.

Most of the older readers of this book likely remember precisely what they were doing on September 11, 2001, when terrorists attacked and brought down WTC buildings in New York. Hardly anyone would know what they were doing on September 11, 2002, or October 11, 2001 (unless you had a great birthday party, got married, split with your spouse, or had some accident). This is because we perceived the terrorist attack as a threat, a dangerous and stressful situation that we should remember.

Similarly, the context of any unpleasant experiences remain embedded in our brains through the action of the amygdala. If we were bitten by a dog or hit by a car, we would remember these events together with all the circumstances around those events for a very long time. This memory is essential, so we recognize the context around the dangerous situation and not only the scary situation. Such remembrance should, at least in theory, help us avoid similar situations in the future.

The amygdala, most likely, has other roles, but these are less understood. The amygdala is probably involved in the processing and memorizing rewarding situations. It likely also has a role in socialization. According to this, people with more extensive social networks seem to have a larger amygdala. Interestingly, the size of the amygdala is different between men and women, and this might be connected with stronger impulsivity in men and often better socialization skills in women.

Another crucial part of the limbic system is the hippocampus, an area of the brain mainly connected with memories. The hippocampus, name meaning seahorse in Greek as the shape of the hippocampus resembles a seahorse, is part of the brain important for the consolidation of memories. Every day, countless events occur to us, and we simply cannot remember everything, as our brains would be overloaded. Therefore, our brains remember some events and not others.

The hippocampus decides what to keep in our memories and prepares the memories for long-term storage. The most crucial evidence for this role of the hippocampus comes from people with a damaged hippocampus. Such people usually retain most of the memories from before their hippocampi became impaired but cannot form any new long-term memories. This kind of amnesia is called anterograde amnesia. It is different from better known retrograde amnesia, where people lose their old memories. Precisely how the hippocampus consolidates memories is not yet understood. The consolidation of memories most likely occurs during sleep, specifically during the active REM phase of the sleep. In this phase of sleep, our brains are very busy, and one of the primary roles of this sleep is likely the consolidation and storage of memories.

People, especially children who do not sleep enough, have poorer memory and worse school performance, as they cannot sufficiently store the memories. Hippocampus also has close connections to the amygdala. Although the amygdala stores the memories of stressful, fearful, and possibly fun events, the hippocampus also plays an essential role in the consolidation of these memories. We strongly remember places where something stressful had occurred to us, and this goes true also in the other way: that we remember places with pleasant experiences. If we were robbed on the street, we would likely get scared every time we walk on that street again. Similarly, a park where we first kissed somebody we loved remains in our memory. Such memory triggers pleasant feelings when we pass such a place again. Memories, consolidated in our hippocampi, are an essential part of love. We want to remember the good times we spent with our partners, and these memories could be crucial when we go through the more turbulent periods of our relationships. Memories of love must be cherished, and the hippocampus has a vital role in this.

The hippocampus has most likely another role, perhaps not directly connected with love, although it might also have some role in sexual reproduction in some animals. The hippocampus helps us to navigate in our environs. The hippocampus, together with another part of the brain in its vicinity, the entorhinal cortex, perceives and analyzes our environments. These two parts of the brain help us navigate through the house, corridors, streets, or woods.

Specialized neural cells called place cells in the hippocampus, and grid cells in the entorhinal cortex, are activated when we walk through our surroundings. They tell us where the obstacles are and where we can move freely. Cells in the hippocampus can also store memories about our spatial surroundings. Cab drivers in London, who have to memorize many streets and the best ways to get to them, have part of the hippocampus, responsible for the spatial memory, larger than people who do not have to remember so many streets. Polygamous meadow voles have larger hippocampi than monogamous prairie voles, presumably, as they have to remember the locations of different females with whom they mate during the mating season. Unfortunately, nobody has yet examined the size of the hippocampi in men who have numerous mistresses and have to remember their addresses.

One crucial part of the limbic system is also the so-called reward system. The main segments of the reward system are parts of the brain called the ventral tegmental area and nucleus accumbens. These areas are activated when we do something good for us and our survival. When we eat tasty food, when we socialize, and when we have sex, cells in the ventral tegmental area release a small molecule called dopamine. Dopamine then travels through the nerve cells into the nucleus accumbens. From there, neural connections go to the prefrontal cortex, and we feel good, satisfied. Indeed, we feel so good that the memories of these good feelings remain in our brain, and we want to repeat such things.

This rewarding feeling is an evolutionary trick. It forces us to do something that helps us survive. Eating a tasty meal gives us energy for survival and thus activates a reward system. Being in a group of friends was always safer in prehistoric times, so socializing also activates the reward system. This system also promotes the survival of the species by providing pleasure during sexual intercourse, which is essential for reproduction and propagation of our genes into the next generation. The system is also activated when we are in love, giving us a pleasant feeling when we are together with the loved one. It is active in mothers when they are in contact with an infant, making this an enjoyable experience that mothers want to repeat.

Unfortunately, this system also has its dark side. The reward system gives us good feelings also when we drink alcohol and take drugs such as heroin, cocaine, and other dangerous drugs. Because the dopamine system gets overstimulated by such drugs, this leads to addiction as the brain becomes accustomed to the overstimulation of the system. After a while, it could get stimulated only with drugs and no longer with normal stimuli that affect non-addicted people such as good food, sex, or just a lovely evening in the company of friends. During addiction, permanent molecular changes occur in the reward system, especially in the nucleus accumbens. It is because of these lasting changes in the structure and function of the neurons in the nucleus accumbens that addictions are so hard to fight and overcome.

Love sometimes feels like an addiction. We do not want to separate from the loved one. When we are separated, we experience signs of anxiety, similar to the withdrawal symptoms in people who are addicted to drugs. Fortunately, this addiction to love is almost always controllable. However, in 3–10% of people, pathological love addiction also occurs. This addiction has very similar symptoms to drug addictions. People with love addiction show uncontrollable, excessive interest toward one (or more) romantic partners. Love addiction ultimately leads to behavioral changes, a lack of self-control, and a renouncement of other interests in life. It could develop into a severe condition with dire consequences. People start to neglect themselves, stop taking care of themselves, and become violent toward the romanticized partner in the most extreme cases.

Fortunately, in most people, the reward system is not overstimulated when we are in love. It just gives us a pleasant feeling of self-fulfillment when we are together with our partner. We also have slight signs of anxiety when we are separated and butterflies in our stomach when we anticipate the reunion with our loved one. But most of us can function reasonably normally, at least after the first surge of love feelings has passed.

Feelings of love are most likely generated in the limbic system or old mammalian brain. Hypothalamus and reward system are responsible for the feelings of attachment, desire to be with the loved one, and emotions of belonging. This might seem strange, as we talk about romantic love as a uniquely human characteristic, yet it is not primarily regulated by parts of the brain that are unique to us, humans. The cerebral cortex, our conscious brain, however, also has a role in romantic love through incessant communication between the cerebral cortex and the limbic system. When we are in love, several parts of the cortex, such as the insula and cingulate cortex, are activated and ceaselessly communicate with the limbic system.

Although falling in love is mostly beyond our control and not our conscious decision, the conscious cortex luckily keeps some control of our primitive brain even when we are madly in love. Unique feelings of love most likely ascend from the interactions between the primitive, old mammalian brain and the conscious human cortex. The prefrontal cortex is less active when we are madly in love, but it is not entirely suspended. We might do crazy things under the influence of the limbic system when we try to win the heart of the loved one. We may neglect the risks and lose the sense of self-protection, but in most cases, our consciousness keeps our primitive brain in rein and keeps us safe even when we fall madly in love.

# Chapter 9
# Why Do Only Male Deer Have Antlers?

*The hope is that laying out what we understand about essential differences in the minds of men and women may lead to greater acceptance and respect of difference.*

Simon Baron-Cohen

On one sunny autumn day, descending after rock climbing in the Alps, I encountered a herd of ibexes. Two majestic males fought, hitting each other hard with gigantic horns. Females stood around, watched the fighting males just as human spectators watch a boxing match. Once the male ibexes noticed my friend and me, they stopped fighting. They looked at us and uttered a loud shriek, utterly unexpected from such a beautiful animal. Ibexes were in a rut, a period for mating when dominant males clash for the females. Although ibexes usually appear



G. Majdic, *Soul Mate Biology*, Fascinating Life Sciences,
https://doi.org/10.1007/978-3-030-67212-6_9

cute and cuddly, the look and sound of these two males with massive horns was nothing but scary, and my friend and I sped away from them. Nature has equipped male ibexes with formidable weapons. These are not used for most of the year but are crucial when the mating season arrives, and male ibexes turn into violent brutes.

Small children of both sexes look very much alike until they go to the beach. When they took off their clothes, boys surprisingly notice that girls are missing something in their private parts. During the young age, in humans and animals, males and females are very similar on the outside, except for sexual organs. However, during puberty, when sex hormones kick in with full steam, sex differences become apparent in many species, including humans. Men grow beards, have a lower voice, are on average taller, and have a more muscular body. Women grew bosoms, are on average shorter, have a higher, more sonorous voice, and usually don't grow beard or hairs on the chest and back. So, in most cases, we can easily identify people as men or women, based on a brief glimpse.

Similarly, in many, but not all, mammalian species, we can easily distinguish between males and females. Male lions have manes around their necks. Bucks have antlers, and mandrills have a colorful face. Male orangutans have chick flaps, and in most mammals, males are larger than females.

Sex chromosomes, carriers of the DNA, determine the sex of the offspring in mammals at the time of conception. While most chromosomes in our cells are present in pairs composed of two very similar partners, sex chromosomes are different. Like all other mammals, we have two dissimilar sex chromosomes, X and Y, responsible for some people being born as girls and some as boys. Girls have sex chromosomes also in pair, two equal Xs. However, males are peculiar creatures and have one X and one Y chromosome in their cells, in addition to 22 pairs of other chromosomes.

Germ cells, eggs and sperm, contain half of the genes of regular cells. During their maturation in testes or ovaries, they undergo a specialized cell division called meiosis, during which chromosomes split in half. Pairs of chromosome separate, so each cell produced during the meiosis has one member from each pair. This is also true for sex chromosomes. Consequently, as all cells in women have two X chromosomes, all eggs contain one X chromosome. However, cells in men (with exception of sperm) contain two different sex chromosomes, X and Y. After meiosis, we therefore have two kinds of sperm, X and Y. This difference in sperm chromosomes determines the sex of the offspring. If the sperm carrying the X chromosome reaches the egg and fuses with it, the baby will be born as a girl. If the winning sperm has the Y chromosome, the baby will be born as a boy.

The Y chromosome is peculiar. It is by far the smallest chromosome in our genome and is slowly degenerating over generations. Some scientists suggest that the Y chromosome might completely disappear in about 4.6 million years from now. But this certainly does not imply that males will evaporate from the face of the earth. Even though sometimes our female partners might want to send us into a black hole, the eventual Y chromosome's disappearance will not help them accomplish that. Not only because it is more than 4 million years away but also because it is unlikely that males will simply disappear. After all, sexual reproduction is crucial for the survival of the species.

The Y chromosome's problem is that it does not have a homolog buddy to hang around with. Y chromosome was initially a homologue of the X chromosome. However, once nature decided that one X chromosome will take a role in determination of male sex, Y chromosome began its separate evolutionary pathways. During millions of years, it became very different from its ancient predecessor, X chromosome.

All other chromosomes, including Xs in female cells, are present in pairs, carrying genes with the same functions. Each gene in our cells is thus present in two carbon copies, apart from the genes on the Y chromosome and genes on the X chromosome in male cells. Therefore, every gene has its backup, just like backups of our computer data on the external hard drives. If one copy of the specific gene is, for whatever reason, damaged or deficient, our cells can counteract the negative consequences of the damaged gene by activating its backup from the other member of the chromosomal pair. And not only that: in the future generations, the damaged gene can be replaced by the working copies of this same gene from intact chromosomes present in other members of the same species, and this is another important role of sexual reproduction.

The Y chromosome, however, does not have its pair. Therefore, if any damages or misspellings in its data occur, this data is lost forever. This loss of information, luckily, does not happen very often. Still, it does happen. During the previous hundreds of millions of years, since the dawn of the mammals, parts of the Y chromosome disappeared completely after becoming dysfunctional, hence the small size of the Y chromosome and a threat it might disappear completely. The X chromosome does not have the same problem. If the male X chromosome loses its data, the damaged X chromosome will be restored in the next generations. Father with a marred X chromosome will likely have a daughter or granddaughter at some point. The second X chromosome from the mother will then eventually serve as a backup to restore the data on the damaged male X chromosome.

Y chromosome might be slowly disappearing, but nature will surely find the way to retain males around, not to annoy women but to sustain sexual reproduction. One way to prevent the complete loss of the Y chromosome is to duplicate its genes, and this is indeed happening. Several genes are present on the Y chromosome in multiple copies. If one copy is lost, other copies remain operative, and the Y chromosome retains its function. So, nature invented the way to create a backup of the data on the same hard drive.

Even if we lose the Y chromosome in 4.6 million years, nature will surely conceive new ways to keep the males. Although women might sometimes think that we are useless creatures, we have at least one important role in nature, bringing half of genetic diversity to our offspring. The ultimate proof that nature will find its way even if the Y chromosome is lost is some mammals in which Y chromosomes already vanished. Yet, these animals still have entirely ordinary males. Two species of rats living in Japan have lost their Y chromosomes during evolution, yet, they have perfectly normal males. It is unknown how exactly sexual determination in these animals is accomplished, but it certainly shows that nature can and will find a way to keep men around, even if the Y chromosome follows the path to extinction.

Although the Y chromosome is crucial in determining the sex of the offspring, it has a relatively minor role in the whole procedure of the development of sexual organs and sexual characteristics. Between the sixth and seventh weeks of gestation in the human fetus, gonads begin to develop. They are initially identical in both males and females. Shortly after the formation of the indifferent gonad, a single gene called SRY, present on the Y chromosome, kicks into action. SRY activity alters the development of indifferent gonad in the male fetus, and gonadal primordium develops into the testis. Once testis is formed, it begins to secrete the well-known, all-powerful male sex hormone testosterone. From then onward, it is testosterone and not the Y chromosome, which is responsible for the development of all males' sex characteristics such as beard, mane, or antlers.

Soon after gonads develop into the testis, testosterone and its close relative dihydrotestosterone trigger the development of male sexual organs: penis, scrotum, epididymis, and sex glands—prostate and seminal vesicles. In contrast to the activity in the male fetus, female sexual organs develop in the absence of hormonal stimulation. Ovaries are inactive before birth and do not assume major hormone-secreting roles until puberty. External sexual organs, penis and scrotum in males and clitoris and vulva in females, develop from the same primordia. In the presence of testosterone, primordium for the clitoris grows into the penis, and the vulva fuses into the scrotum. In the absence of testosterone, the clitoris remains small, and the vulva remains open, creating an entrance into the vagina.

Womb, vagina, and external sex organs in baby girls, therefore, develop without the need for ovarian hormones during embryonic life. Consequently, in the complete absence of gonads (either testes or ovaries), which occurs because of some rare mutations, the baby develops with female genitalia, even if its cells contain the Y chromosome. So, the default developmental pathway in humans is the development of the female genitals, and development of male sexual organs is a deviation from the "normal" pathway.

Testosterone's importance in male development is perhaps most clearly shown in a rare disease called testicular feminization syndrome. Patients with this disease have an inactive receptor for testosterone. Although they have testes in the abdomen and produce testosterone, they are born with females' external sexual organs. Testes remain in the abdomen, the scrotum is not formed, and the penis remains as a clitoris. However, as testes also secrete a hormone that triggers the regression of womb and oviducts in female fetuses, patients with this disease do not have wombs and oviducts. They are often not diagnosed until puberty when such girls do not get periods, as they do not have ovaries. Interestingly, their breasts usually develop, as estrogens are produced from the copious amounts of testosterone in their bodies and can act through estrogen receptors that are not mutated.

Another somewhat similar disease is caused by a mutation in a gene that transforms testosterone into a more powerful dihydrotestosterone. This disease is even more peculiar. Dihydrotestosterone is responsible for the full development of the penis and scrotum in a baby before birth. Therefore, babies lacking this hormone are born as girls.

However, during puberty, the amounts of testosterone increase to much higher levels than before birth, and external sexual organs become more sensitive to regular testosterone. As a consequence, these girls seem to switch sex during puberty. Their clitorises become enlarged to resemble a small penis. Such children acquire deep voice, vaginal labia grow and become similar to the scrotum, and patients develop a more muscular body. Essentially, girls seem to be turning into boys. This is a complicated but luckily very rare disease that occurs more often in some parts of the world. It is prevalent in some regions of the Dominican Republic, where they have a unique name for this genetic disease—guevedoces, literally meaning "penis at 12."

Interestingly, the facial hair in "guevedoces" patients is usually less prominent than in healthy men, as hair growth remains dependent on the dihydrotestosterone throughout life. For the same reason, such patients do not become bald when they are older, as dihydrotestosterone causes baldness. However, before somebody makes a hasty conclusion, baldness is also dependent on the genetic predisposition. Therefore, having a bald head does not mean that such men have more testosterone and are more masculine than men with a thick mane on their heads. It only means bald men have inherited the tendency to be bald from their fathers and mothers.

Once a healthy baby boy or baby girl is born, they remain externally quite similar until puberty. Immediately after birth, testes in male boys become sedentary, and both testes and ovaries remain mostly inactive from birth until puberty when hormones suddenly flood teenagers' bodies. The hypothalamus and the pituitary send signals to gonads to start producing sex hormones in large quantities. This flood of sex hormones affects the whole body, causes the development of typical sexual characteristics, and influences the brain. Testosterone and its relatives have a principal role in the development of male sexual features. In females, ovaries finally become active during puberty. They begin to secrete two groups of female sex hormones: estrogens and progestins. The most potent estrogen, estradiol, is the hormone responsible for the development of female secondary sexual characteristics such as different distribution of the body fat, thicker and smoother skin, and growth of the bosom.

During puberty, testosterone and estradiol, together with the hormones from the pituitary, also activate gonads. Consequently, sperm are now produced in the testis, eggs start to mature in the ovaries, and girls begin to have menses. With these bodily changes, naturally, the interest in the opposite sex grows. Suddenly, girls are no longer horrible, sensitive, crying, and annoying creatures for boys. Similarly, boys are no longer dirty, violent misbehaving monsters for girls. With all these hormonal changes during puberty, the bodies change, and love starts to arise within the teenagers' bodies and brains.

In puberty, sex differences between men and women become apparent, even though we are not the most sexually dimorphic animals. Men grow beards and acquire a lower voice. They often have a more muscular body, although this is frequently replaced by a beer belly in more mature years, when another male-specific natural gift occurs—baldness. Androgens also influence these not-so-desirable male characteristics, which is why both beer belly and baldness are usually found only in men.

If we look at animals, differences between sexes are often much more remarkable. Most of us have seen deer or moose antlers, which can grow to impressive size. Male gorillas and orangutans can be twice the size of the females, while elephant seal males can weigh up to four times as much as females. Lion manes, together with low-pitched, loud roaring, earned lions the entitlement of "the king of the animals." In birds, there are many beautiful examples of sexual dimorphism, either in colorful plumage in ducks and pheasants, large peacock tails, or beautiful singing by songbirds, often performed only by males.

As in humans, in other mammals and birds, sex differences are instigated by testosterone. Testosterone stimulates deer antlers' annual growth, and antlers fall off when testosterone levels decline at the end of the reproductive season. Male bird singing is under the direct influence of androgens, and song brain centers were extensively studied in birds to understand how androgens influence brain development and function. Androgens directly induce lion mane growth. Castrated lions lose the mane. A few years ago, a lioness grew the mane in a South African zoo. Veterinarians found out that she had elevated levels of androgens produced by her ovaries. When she was cured, and her androgen levels dropped, she lost her mane as well.

But why have such sexual dimorphisms developed? This question has intrigued scientists for a long time. Notwithstanding, one of the greatest biologists of all time, Charles Darwin, wrote in 1860: "The sight of a feather in a peacock tail, whenever I gaze at it, makes me sick." He was troubled by how to explain sexual selection in view of his evolutionary theory. Large, seemingly useless, ornaments are not beneficial for the survival of the individual. Large antlers hamper deer walking in the thick woods. Colorful plumage in male birds alerts predators. To solve this question, Darwin suggested that females choosing their mates were the real drivers of the evolution of male ornaments. He correctly assumed that females prefer large antlers, thick mane, and colorful feathers, and this drives the evolution of such ornaments to sometimes absurd appendages.

In many mammals, albeit with a considerable number of exceptions, males try to mate with as many females as possible, sowing their wild oats on many fields. Sperm are a dime a dozen, easy to come by, and bargain to throw around. Eggs, however, are a much more precious commodity. Only one or at most few are produced during every reproductive cycle, which is in many wild animals only once a year. To preserve female genes, mammalian mothers must take great care of their younglings once they are born. This requires a lot of investment in time and energy, so a female must choose the best mate: one who will provide the best genes for her brood and thus ensure the survival of the next generation.

Most females cannot do genetic analyses. Even if they would be able to, knowing the sequence of your genome does not answer whose genes are better. Females must, therefore, rely on other signs to determine who might be the best father. In birds, these signs are colorful plumage; in deer and moose, large antlers; in lion, grandiose mane; in orangutan, enormous chick flaps; and in many mammals, just the body size. The larger the male, most likely the healthier, and with the best genes. Therefore, the choice for females is obvious. Choose the moose with the most massive antlers. He must be healthy and strong if he can survive in the forest with these giant awkward ornaments on his head. Pick the most muscular elephant seal. He is undoubtedly healthy if he was able to build such body mass and defeat all the opponents. Go with the lion with the thickest mane. He has plenty of testosterone, which gives him bushy mane and strengthened muscles, needed to survive in the harsh African savannah.

It is fussy females who drive the evolution of the male ornaments to sometimes absurd situations when ornaments could be troublesome for males. Yet, males continue to compete with each other for the ultimate prize, a chance to mate with a female. As a consequence of this race, males grow appendages, muscles, or sometimes even brain to be more successful in this ever-present competition.

Although sexual dimorphism is not as prominent in humans, with men being only 10–15% larger than women, we use various ornaments to attract the opposite sex. Just like deer is flashing his antlers, so are human males, full of testosterone and ready to mate, flashing their fast sports cars in front of the picky females. But this has more to do with the lust and desire to reproduce than with true love. Yes, Snow White has fallen for the prince on the white stallion, but long-lasting love does not require flashing cars or expensive holidays on tropical islands. Love is found in all layers of society. Neither the most muscular men nor the most beautiful women are the only ones who can find their soulmates. After all, Belle fell in love with the fearful but gentle beast and not with the handsome but brute Gaston in the Beauty and the Beast movie.

Indeed, it is not the size of the antlers or the mane's thickness that is important for a long-lasting relationship. In animals where males flash their large ornaments or just brag about their body size, males are not faithful and usually only look for a one-night stand. In mammals, sex differences are the smallest in animals that live together in seemingly harmonious relationships. Prairie voles, dik-dik antelopes, or gibbons are all animals that live in monogamous relationships. In all, we would struggle to tell who is male and who is female. So, the small difference between sexes in our species might tell us something. Perhaps we were made to love, stay together, and care in unison for our brood. It may suggest that we were not created to live a solitary life of single moms struggling to up-bring their children and selfish bachelors looking to sow their oats around. Nevertheless, we are a very adaptable species that is able to survive in various environments and in different circumstances. So, what I said above does not mean that it is wrong if somebody prefer solitary life. But we are in general social beings. Most of us long to be with somebody, and this might be evolutionary embedded into our genes.

When we talk about sex differences in us humans, most of us think about sexual organs. However, these are not the only differences between male and female bodies. Although we often believe that our bodies, except for reproductive organs, function in the same way, we know today that this is far from the truth. Differences between males and females exist in many internal organs, often in their structure and function.

Organs and organ systems such as the guts, kidneys, muscles, liver, immune system, heart, blood vessels, and others all differ between males and females, with significant consequences for the functioning of our body in health and disease. For example, sex differences in the functioning of the immune system and probably in better vascularization of the skin contribute to better healing of wounds in women. Yet, these same differences in the immune system cause more autoimmune diseases in women, such as lupus, rheumatoid arthritis, thyroid diseases, celiac disease, and multiple sclerosis.

Women are less often inflicted with heart problems. The incidence of cardiovascular diseases and deaths from such causes is lower in women, especially in premenopausal women. This difference in cardiovascular diseases is the direct effect of estrogens. After menopause, when ovaries stop producing estrogens, the incidence of these diseases in women starts to increase. But all is not good and well even here. Signs of cardiac arrest differ between men and women. As women are less likely to suffer from heart failure, physicians sometimes overlook symptoms of cardiac arrest in women, as they are used on symptoms in male patients.

Sex differences in the liver are especially important because toxins and drugs are metabolized and eliminated differently from the male and female bodies. Thus, some drugs act differently in men and women and should be taken differently. There are differences in energy metabolism between sexes and many others. Consequently, it is essential to keep in mind that biologically, men and women are not the same. Although we should and must have equal rights and equal opportunities, there are biological differences between our bodies. These should not be neglected as they are important in health, and especially in disease!

Evolutionary, sex differences in various body organs are more difficult to explain than differences in appearance, which can be directly correlated to the courtship and mating. Thus, differences in internal organs more likely arose from dissimilar behavior between males and females in the animal world. Suppose a female stays in her burrow and its vicinity and wait for a male to find her. In such a case, it makes sense that males have better spatial orientation than females of such species. It is males who must search around and remember possible locations of the willing females. Mammalian females have very high and specific energy requirements during pregnancy. Thus, it is understandable that the liver, an organ intimately involved in the energy partitioning in the body, functions differently in men and women.

Receptors for sex hormones, estrogens and androgens, are also abundant in the brain. Although sex differences in the brain are often hotly debated, numerous studies have shown that sex differences in both structure and function of the brain exist. These sex differences have important implications not just for everyday life but also for psychiatric diseases. Many mental disorders occur preferentially in one sex or have different signs in men or women.

Sex differences are present in the autonomous brain, hypothalamus, amygdala, reward pathway, and others. Most sex differences are associated with sex hormones, which can act on the brain in two ways. During development in the womb and later during puberty, sex hormones induce structural differences in the brain that persist throughout life. These are called organizational differences, as the structure of the brain is changed (organized) permanently. In adult life, sex hormones influence specific brain areas' activity, and these effects are therefore called activational effects of hormones. In recent years, several studies have shown that genes on the X and Y chromosomes also cause sex differences in the brain. Some sex differences induced by the hormones might not exist to produce variations in the behavior but actually to mask structural differences due to sex chromosomes and influence more similar brain functioning between sexes.

Some of the most convincing sex differences in human behavior are spatial abilities, verbal communication, and empathy. Women usually have better verbal skills and show much higher levels of empathy. On the other hand, animal and human males often have better spatial skills than females. Better spatial abilities are correlated with the size of the hippocampus, part of the brain with a role in spatial orientation. Hippocampus is often larger in males, hence better spatial orientation in this sex. But with all sex differences in the brain, it is essential to remember that these differences are not absolute. Not all men excel in spatial orientation, and not all women are better than men in verbal skills. This is true for most sex differences both in the behavior and in the structure of the brain. Sex differences in the brain are usually present in a large variety in both males and females. When we look at large groups of men and women, we can see that the differences exist. On average, men are better at certain things, and women are better at others, but in individual men and women, these differences might not be present.

When studying sex differences in the brain, we always have to consider also social influences. The brain is a very adaptable organ, and genes, hormones, and experiences throughout life can change its structure. Therefore, it is difficult to conclude whether boys prefer to play with cars and girls with dolls because their brains are hardwired for this or because we encourage the boys to play with cars and girls to play with dolls. However, in this particular question, we do have some suggestions that toy preferences are, to some extent, embedded in our brains. When infant monkeys were given different toys to play with, monkey girls preferred plush animals, and monkey boys liked toys with wheels, very similar to human children.

The relative size of the sex difference in the brain sometimes correlates with the social habits of animals. Polygamous meadow vole males have better spatial orientation and larger hippocampus compared to the monogamous prairie vole males. In meadow vole males, spatial abilities and size of the hippocampus are also larger in males than in females. In contrast, in prairie voles, there is no noticeable sex difference in the size of the hippocampus. This difference is likely associated with their mating habits. Prairie vole males stay with their mate after the first sexual encounter. They do not have to explore a large area searching for willing females. Meadow vole males, however, are ever searching for new adventures. Better spatial orientation due to the larger hippocampus might help them to be more successful in locating burrows of all willing ladies in the given territory. Like London cab drivers who have larger hippocampi, so they can find all the streets in the city, meadow voles might have grown larger hippocampi so they can find girls, scattered in their burrows.

Visible differences between sexes are widespread in the animal world. In mammals, they range from very prominent to almost non-existent. Correlations between mating habits, courtship behavior, and sex differences indicate that more promiscuous mammals display more obvious sex differences, either in body size or in the presence of ornaments such as antlers or manes. This discrepancy is usually explained by the evolutionary pressure on males to fight between themselves to capture the ladies' hearts or by choosy females who want the largest and the most beautiful mates as fathers for their brood. In animals with males competing for the ladies' love, males contest each other with strength and beautifulness. They fight for the ultimate prize—a chance to mate. But this has nothing to do with love. Animals that have the fiercest fights and display the most elaborate ornaments to win the right to mate do not cohabitate in harmonious relationships. Males abandon their partners as soon as they finish their business. Such males seek new partners again and again and are never faithful to one female. They procreate but never look back over their shoulder to check on their sexual partners or their infants, let alone to help moms raise and protect the children.

In animals that mate only with one partner and males stay with this partner for life, there is less competition between males as a struggle for females is not recurring every season. The majority of males have a chance to find a partner. Sex differences in monogamous animals are, therefore, much less prominent. By judging the extent of differences between males and females, we can usually make a very good guess whether the animals are monogamous or polygamous. So, it seems that love, attachment, and long-lasting relationships are inversely correlated with the biological differences between males and females.

Although we look different, and our bodies function differently, sex differences between men and women are not as prominent as in numerous animals. Therefore, this might suggest that we were created for love. We are meant to stay together in lovable relationships, care for our partners and our children, and not perpetually seek greener pastures or fulfillment of natural instincts.

# Chapter 10
# Of Bees and Flowers

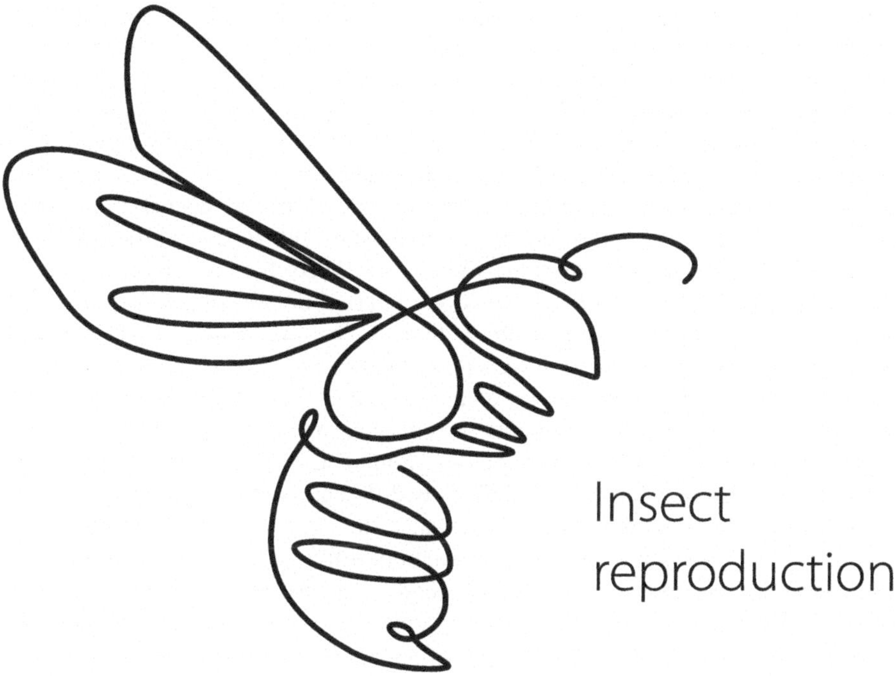


*Ask yourself my love whether you are not very cruel to have so entrammelled me, so destroyed my freedom. Will you confess this in the Letter you must write immediately, and do all you can to console me in it—make it rich as a draught of poppies to intoxicate me—write the softest words and kiss them that I may at least touch my lips where yours have been. For myself I know not how to express my devotion to so fair a form: I want a brighter word than bright, a fairer word than fair. I almost wish we were butterflies and liv'd but three summer days—three such days with you I could fill with more delight than fifty common years could ever contain.*

John Keats

My father was a beekeeper, and as a kid, I often helped him. I learned that bees do not sting out of evil, but only when they feel threatened. I also learned how to distinguish worker bee, drone, and a queen. At that time, I was not concerned by the peculiarity of having three types of bees instead of only two, what is more common

 
G. Majdic, *Soul Mate Biology*, Fascinating Life Sciences,
https://doi.org/10.1007/978-3-030-67212-6_10

in nature. I learned how to recognize the queen and was always excited when I spotted the queen in the hive.

However, Aristotle, one of the most famous Greek philosophers, a keen observer of nature, was troubled by the bees. He wrote more than 2300 years ago: "The generation of bees is beset with many problems. If there are three classes and two sexes, something is amiss." Aristotle correctly noticed and described that in bees, something funny is going on regarding their sexes. While in most animals there are two sexes that are often clearly distinguishable, there are three kinds of individuals in bees: workers, drones, and queens. Aristotle thought queens were kings, based on the observation that bee queens have stings and drones have not. In his opinion, only males carry a weapon in nature, and therefore, bee queens must surely be males.

He correctly noticed peculiarity regarding sexes in bees. In almost all other organisms, there is either one kind (in organisms with parthenogenetic reproduction) or two kinds (males and females in organisms with sexual reproduction) of individuals, but not in bees. Aristotle rightly concluded that one group of bees is sterile, as indeed are worker bees. However, he erred in believing that drones and queens reproduce parthenogenetically, without sexual intercourse. This view was based on his observations as he never saw copulation in bees and neither have beekeepers he talked with. Nowadays, we know he was wrong. We know that not only bees but also a vast majority of insects reproduce sexually.

The insects are by far the most populous group of animals on earth. As such, it is not surprising that they have developed many different ways of reproduction.

Although some insects reproduce asexually, the vast majority of insects reproduce sexually. Yet, even in sexual reproduction, different species utilize different approaches. Some insects reproduce by copulation, just like mammals. In many others, fertilization occurs without close contact between males and females. In insects without intercourse, males deposit their sperm in special sacs, called spermatophores, in the absence of the females. Females must find and pick up the spermatophore to become impregnated. The deposition of these sacs full of sperm is likely the switching point between reproduction in the aquatic environment and reproduction in terrestrial animals. In the marine environment, sperm and eggs can be simply deposited in a safe place, what many fishes still do. There is no need for any special protection, such as spermatophore, as eggs and sperm are stable in the aquatic environment. Outside of water, however, cells like sperm and eggs quickly desiccate and die. Therefore, cells must be protected from drying out. Protection is provided by wrapping the sperm in a special protein coating, called a spermatophore.

Deposited spermatophore, essentially a sack full of sperm, is picked up by the female. Many female insects store the sperm in their bodies for a very long time, quite often for the rest of their life spans. Some primitive insect males indeed leave spermatophores simply lying around on the ground or on plants, leaving them for females to find them and pick them up randomly. But evolution quickly determined that this is not the most efficient process to ensure the mating and continuation of the species. Females might not find spermatophores lying indiscriminately around. Predators could eat spermatophores as they provide a very nutritional snack. The force of nature, such as heavy rain, could swipe the spermatophore away. Therefore, some insects developed a procedure in which male insects still deposit the spermatophore on the ground or on a plant but then attempt to attract females. Insect bloke lures the female into the vicinity of a deposited spermatophore. Essentially, he is performing some kind of primitive courtship, but without ending in copulation. Thus, the male ensures that the spermatophore is found and picked up by the gal.

But this was just an intermediate step in evolution. To ensure even more efficient reproduction, insects soon developed internal copulation, a process similar to mating in mammals. In insects with internal copulation, males often produce a spermatophore, a protein sac full of sperm. Yet, males deposit spermatophore directly into the female genital tract, and females can then use the sperm as needed. In other insects, such as honeybees, males got rid of spermatophores and now deposit their semen directly into the female genital tract, without special wrapping. However, females still store these sperm for a long time. During copulation, the drone deposits his sperm in a special sac inside the bee queen called the spermatheca. Spermatheca cells then nourish and watch over the sperm, ensuring that sperm can survive inside the female reproductive tract for a long time.

Unlike mammals and birds, insects usually copulate only once in their lifetime. Males are needed only once, and therefore, males are dispensable after copulation. Indeed, in many insects, males die soon after mating, as they have fulfilled their mission. In some species, they not only die but are devoured by their loved ones. In praying mantis and some other insects, and even more often in spiders (who are not insects!), a female eats her mate, or sometimes only parts of his body. Some cricket females eat their partners' wings after the mating but let them live winglessly. It is a bit like your fiancé would chop your legs off after the first night together, not a very appealing prospect. And it is not even that the ladies want to keep their mates close by and prevent them from flying away by eating their wings! Males are free to leave without their wings, and females seek new partners later, preferably virgins with the intact wings.

Sacrificing your wings is probably better than being a mate of a praying mantis female who eats her partner immediately after copulation. Sometimes she begins feasting on the male while the mating is still going on. This behavior likely seems repulsive. Still, it has a perfect sense from an evolutionary perspective for both males and females. Male is needed only once for copulation and is not required to provide parental care for the offspring. However, his body can provide an excellent resource of foodstuffs needed by the female to ensure the survival of the offspring. Males sacrifice themselves for the greater benefit. They give the female and her offspring a better chance of survival, thus providing a better opportunity for their genes to continue their evolutionary path from one generation to another.

Yet, males do not always reconcile with the evolutionary grand scheme of things and fight for their survival. Especially spiders developed some exciting ways to avoid being eaten. Male orb-web spiders opportunistically wait until the females are hungry or distracted by something else. Once the lady is not looking, they commence the mating with the startled female. In some orb-web spiders, males even bring some food as a nuptial gift. While the female is consuming the present, the male finishes with the intercourse, thus reducing the risk of ending as a tasty meal.

Some mantis males make sure they approach the female in very windy weather, reducing the chances of the female hearing them. Desert grass spiders use a dirty trick, akin to what would be considered a criminal activity in our societies. They use pheromones as date-rape drugs. When a male approaches the female, it releases unique pheromones that induce deep sleep, almost a coma in the female spider. It can then copulate with the sleeping beauty without risking to be devoured.

Although only once, males are necessary for the continuation of the species. Therefore, two sexes must be present in the population. In insects, two sexes are obtained in different ways in different species. In some insects, it is quite similar to mammals, with X and Y chromosomes being responsible for determining who will be male and who will be female. In mammals, the Y chromosome harboring sex-determining gene SRY is the master switch, ushering the fetus's development from the default female pathway toward the male phenotype.

In insects, however, it could be either the Y chromosome a switch forcing the development of a male or the X chromosome a switch forcing female sex development. In mammals, the presence of the Y chromosome always triggers the development of a male, even when there is a surplus of X chromosomes. Individuals with two X and one Y chromosomes develop as males. In some insects, similar chromosomal situations (XXY) will instigate a male's development and in others development of a female. So, unlike in mammals, there is a large variety in how chromosomes determine insects' sex. Furthermore, XY chromosomes are not the only sex-determining mechanism in insects. Some insects use a similar system as birds. Females have two different sex chromosomes (Z and W) and males two identical sex chromosomes (ZZ). In some other insects, the number of X chromosomes in cells is a switch triggering female or male development.

Such an example is honeybees. Male and female cells have only X chromosomes, and it is the number of Xs that determines whether somebody is a boy or a girl. Females always have two X chromosomes (both workers and queens) as they develop from fertilized eggs (eggs fusing with sperm). In contrast, males develop from unfertilized eggs and have only one X chromosome (from the egg). In honeybees, utilizing this sex-determination system, the queen can decide whether her offspring will be male or female. After mating and storing the sperm in the spermatheca, the queen can choose whether to fertilize the eggs with the stored sperm or not. If she fertilizes her eggs, the offspring developing from these eggs will be females. Most of them workers, while queens develop from the same fertilized eggs if they receive special diet made of royal jelly. On the contrary, if the queen decides not to fertilize the eggs, all progeny from such eggs will be males (drones), an accomplishment many prospective human parents long for but are unable to achieve while it comes naturally to the honeybees.

Honeybees mate only in the air. When the space in the hive or the tree becomes too tight, an old queen with half of the bees swarm away and finds a new home. The young queen takes over the hive, and she firstly flies out of the hive to meet drones, prospective partners. Although honeybee queens are not as gruesome as praying mantis and do not chew their partners, copulation is still mortally dangerous for the drones. In an attempt to ensure that only his sperm will impregnate the female, the drone's phallus is ripped off after the mating to close the queen's genitals. However, with an open abdomen resulting from phallus sequestration, drones invariably die after mating. Yet, the ultimate sacrifice is not wholly successful. After the mating, another drone can remove the previous mate's phallus, and the queen often mates with up to ten drones during one mating bout.

All beekeepers know that sometimes the queen begin to lay eggs without swarming outside the hive and without being fertilized by drones. Such an unimpregnated queen will produce offspring, but only males with one sex chromosome. She does not have sperm to fertilize the eggs. Such a colony with an unimpregnated queen could continue to reproduce for one generation but dies out sooner or later, as only males are born.

Mating in insects is an elaborate process that involves copulation and often quite comprehensive rituals aimed to attract the best possible partner. Females who are willing to mate usually send out pheromonal signals to attract males. Such pheromones have crucial roles in insect reproduction. Joseph Lintner, an American entomologist who observed silk moth males aggregating outside his office window, correctly assumed that some chemicals produced by female moths attract male moths. We know today that, indeed, most insects secrete pheromones that help male and female insects attract and recognize each other. Some insects detect pheromones in really minuscule amounts. Certain male insects can detect pheromones secreted by females up to 10 km away! The pheromones in the air lure males toward females, and males try to mate with females once aroused by pheromones.

Although pheromones are the most common way of communication between male and female insects, it is not the only one. Some insects like mosquitos use sounds instead of pheromones, just like men used to sing under the windows of their beloved ones. Unfortunately, mosquito love songs are not adjusted to our ears. When we hear them, they are a nuisance, rather than the remainder of love. But perhaps we should not be too judgmental. Our love songs may be similarly annoying to mosquitos, as mosquitos' buzzing annoys us.

Other insects such as fireflies use light signals emitted by specialized luminescent organs in the search for love. Fireflies flying around on the midsummer night are not doing this to please us on warm summer nights but to communicate between themselves. Firefly boys fly around, blink their light, searching for girls. At the same time, gals reside on the plants in the darkness and wait for a suitable suitor. Only if the passing male excites them with his light females start to blink their light to signal the male that they are there and ready to mate. This pattern is a simple way to distinguish male and female fireflies—the ones flying around and blinking with their lights are males. Those sitting on the plants, mostly in the dark and flashing only occasionally, are females.

However, animals are smart, and some fireflies have evolved to use blinking for their advantage. Some predatory fireflies developed a blinking similar to the blinking of other fireflies, their prey. Hungry predator firefly sits on a plant and blinks a female signal to naïve, horny males seeking a mate. Unfortunately for them, when they approach the light of what they presume is a willing female, a predatory firefly waits for them. Instead of being sexually satisfied, they are gobbled by the blinking cheating predator.

Once the insect male finds and recognizes the female, it is often not enough to start mating. In many insects, males have to impress females. As insects do not drive expensive, flashy cars or take their beloved to great restaurants, insect males developed other ways to impress females. Gals are often fussy and not willing to mate with any male trying to do so. In different species, males use different approaches to win, whatever serves as a figurative heart in insects. Males perform ritual dancing, show off with their colors, or try to impress the females by the size of their antennas. Sometimes they show off by flying performances, gently stroke the females by their legs or antennas, and similar. Only if they sufficiently impress the female will she allow the male to copulate with her.

Water strider males use a very selfish approach to force the unwilling female into mating. Water strider females are, like many other insect females, quite choosy. They usually stay close to the surface of the water, unwilling to mate with every male approaching. However, an undesired water strider suitor will sit on the top of the female and make small ripples on the water's surface, attracting the fish. As the female is closer to the water's surface, she is more in danger of being eaten by the fish. Male thus blackmails the female—either she will give in and mate with him or he will attract the predatory fish to eat the female water strider.

Although most insects have specialized organs like the penis and vagina, bedbug females do not possess a particular organ for copulation. During the mating, a male bedbug pierces the outer layer of the female body and deposits his sperm inside her darling. These bugs, however, are not very smart, but they seem to be quite horny. Males are not pursuing and mating only with females but also with other males as male and female bedbugs look pretty similar. Unfortunately for male bedbugs that are being mistaken for females, this piercing through their outer layer causes them to die. Males have not evolved a unique mechanism, as females did, to protect them from the harmful effects of being pierced by the penis. Therefore, for a male bedbug that mistakenly mates with a male, this might not be only a mistake but also a victory in the mating wars. He gets rid of his competitors in the everlasting race of who will transfer its genes into the next generation.

Insects do not write poetry or drive flashy cars. Yet, males fight fiercely to win the females' hearts. Unlike in humans, when we usually lose our heads only figuratively when in love, in insects, this often happens in reality. Yet, even though the mating presents peril for many insect males, males fight for love. Males put all their effort and compete with other males to be chosen as partners of the beloved females, just as humans do, even though the mating is also their demise. Although the mating between male and female insects cannot be called love, there are some similarities in courtship between insects and mammals. Therefore, some origins of the behavior that later developed into the love most likely exist even in such groups of animals like insects that have seemingly nothing in common with us.

# Chapter 11
# After-Dark Secrets of Nemo's Parents

Fish
mating
and

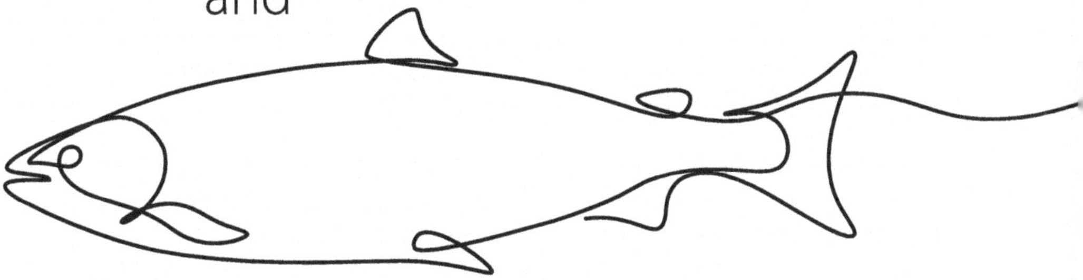

caring
for
infants

*Of course I like you. It's because I like you I don't want to be with you. It's a complicated emotion.*

Marlin in Finding Nemo

Most of us remember Nemo, a cute clownfish from the cartoon Finding Nemo. At the beginning of the movie, expecting clownfish parents Coral and Marlin, obviously in love with each other, prepare the nest for their offspring. After barracuda eats Coral and most of her progeny, Marlin remains alone with his only son Nemo and becomes a very protective parent. When a scuba diver abducts Nemo, loving Nemo's father sets on a perilous search through the ocean, determined to find and rescue his son.

Perhaps Finding Nemo was so popular because it showed us fish from a different perspective from what we usually think about them. Fish are generally not seen as caring, loving, monogamous, and parental animals. We surely do not associate the fish with love, as we do with some other animals, such as swans. Nevertheless,



G. Majdic, *Soul Mate Biology*, Fascinating Life Sciences,
https://doi.org/10.1007/978-3-030-67212-6_11

several species of fishes are monogamous, and many fish parents take good care of their young.

Fishes are usually considered as the most primitive vertebrates (animals with spines). They are a very diverse group of animals, and the total number of fish species is unknown. It is estimated to be at least 22,000, but possibly over 30,000 what is more than all other vertebrates: reptiles, amphibia, birds, and mammals, combined. As fishes are a very varied group of animals, they also have very diverse ways of reproducing. Some fishes can reproduce parthenogenetically, without two sexes, with a virgin birth, what a direct translation of the Greek word parthenogenesis means. At least 20 species of fishes can reproduce asexually and have thus individuals of only one sex. A single egg that develops inside the female ovary grows into the adult fish without receiving half of the genetic material from a male. Some parthenogenetic fishes reproduce all the time asexually, while some females, including hammerhead and blacktip sharks, can reproduce parthenogenetically if there are no males around.

The Amazon molly, a fish native to northern Mexico and Texas, not to the Amazon river as the name suggests, performs unique parthenogenesis. These fish are named after the Amazon female warriors from Greek mythology due to their peculiar reproductive process. Amazons were fierce female fighters that lived in the exclusive women-only settlements. According to Greek mythology, they met once a year with males from the nearby tribe of Gargareans just to have sexual intercourse, become pregnant, and obtain the next generation's offspring. Like Amazon warriors who were all ladies, all Amazon mollies are females. Their eggs do not need to receive genetic material from the sperm to fully develop, unlike Amazon fighters who needed to be impregnated by Gargarean males. But, as in the animals that reproduce sexually, Amazon molly eggs need a trigger to begin developing into fish. This trigger is close contact with sperm.

As Amazon mollies are asexual, with all individuals being females, there are no males to mate with and no sperm to trigger the eggs' development. To solve this problem, Amazon mollies borrow the sperm from males of the related fish species that live in the same area. Just as Amazons from Greek mythology met with foreign adult males to mate with and obtain offspring, Amazon mollies lure Atlantic molly, sailfin molly, or broadspotted molly males into the mating. But strangely, Amazon mollies must have sexual intercourse with males. They do not merely release their eggs into the water and leave them to be found by the sperm, as many fishes do. As there are no Amazon molly males to mate, they must entice males of other species into the copulation. How do they convince males and why males from different species mate with Amazon molly is not known.

Mating between different species is very rare in the animal world, and it is against the evolutionary sense. Males mating with females of other species would spend their time and energy on something (sexual intercourse) that is entirely worthless. It will neither help them survive nor promote the survival of their species. However, it seems that males who mate with Amazon molly females have more success with females of their species. Perhaps mating with Amazon molly female hints to the females from their own species that such males are good at reproducing and thus worthy mates to succumb to.

Sexual reproduction is beneficial for the survival of any species. Therefore, it is difficult to understand why some fishes use parthenogenesis. Transient parthenogenesis, as observed in sharks, could undoubtedly be beneficial as females will reproduce even in the absence of males. Females could, therefore, take care of the continuation of the species when males are absent, while genetic diversity is reintroduced when parthenogenetic offspring mate in the next generation and continue to reproduce sexually. More curious is the asexual reproduction in fish such as Amazon molly. Genetic analyses have shown that these fish reproduce strictly asexually for at least 100,000 years. Yet, they do not seem to amass harmful mutations, and they thrive in their environment. How do Amazon mollies prevent the accumulation of deadly mutations in their genomes remains one of the questions for science in the future.

These parthenogenetic habits are a peculiarity in the realm of fishes, and most fishes do reproduce sexually. For this, a species needs to have two sexes, and the processes of determining who is male and female are extremely varied in fish. In some species, similar mechanisms as in insects, birds, and mammals, using sex chromosomes, are utilized. Both XY (males having XY chromosomes and females two X chromosomes) system, similar to mammals, and ZW (males having two Z chromosomes and females having ZW chromosomes) system, similar to birds, are used in fishes.

However, in many fishes, other factors trigger sex differentiation, in some, the presence of certain hormones, similar to human sex hormones. In others, the water temperature or the number of fish living in the same area determines the sex. In yet another fishes, sex is determined by the amount of oxygen in the water or the water's acidity. But in general, sex in fish is not as strictly defined as in mammals and birds, where sex is determined conclusively at conception and is constant until death. While gonads in mammals and birds are largely developed at the time of birth, the development of gonads lasts much longer in fishes. The gonads could differentiate into testis or ovary quite late when the fish is already fully grown.

Furthermore, the number of fish species is bisexual, with both testes and ovaries, a condition called hermaphroditism. The term hermaphrodite, which is also used for certain infrequent human conditions, when babies are born with gonads of both sexes, arose from Greek mythology. Hermaphroditus was a gorgeous son of Greek gods Hermes and Aphrodite. A nymph called Salmacis fell in love with Hermaphroditus and asked gods to unite them forever. Gods granted her wish more thoroughly than she probably desired. They merged them into a single being that had characteristics of both man and woman. In remembrance of Hermaphroditus, the word hermaphrodite is still in use today for animals or humans with characteristics of both sexes.

Some fishes are real hermaphrodites, possessing both testicular and ovarian tissues at the same time. Much more common, however, is the so-called transient or sequential hermaphroditism. Individual fish is at a particular time of its life a female, producing eggs in the ovary. Later, it switches sex, develops testes, and starts to produce sperm, or vice versa. Although the Finding Nemo cartoon was cute, it did not depict the habits of clownfish accurately. In one sea anemone, which is indeed the home of the clownfish, two larger and several smaller individuals usually live together. Two larger clownfish are sexually mature male and female, and all small fish are sexually immature males. If the large female dies, the adult male changes sex and converts into a female. One of the little males (usually the largest) develops into the sexually mature male and mate with a newly sexually matured female. In clownfish, the change of sex is always in this direction—males changing into females, which is called protandry.

The opposite is females transitioning into males. This change is called protogyny, is more common than protandry, and is present in numerous fishes. It is usually occurring in social fishes where one dominant male mates with many females and protects both his territory and females aggressively from other males. For other fish, potentially young males, it makes sense to live as females and reproduce as females. Such females can pass their genes onward while they are young and small, what they wouldn't be able to do if they would be submissive males, unable to mate. Once they grow enough to challenge the dominant male, or a dominant male dies, one female converts into a male and begins to mate with his female friends. This transition is common in numerous wrasses, groupers, parrotfishes, gobies, emperors, and other, mostly tropical fishes.

Less common is bidirectional hermaphroditism, where individuals can change sex in either direction more than once throughout their lifetime. In Indo-pacific cleaner wrasse and some other fish, school typically consists of one large male and several small female fish, as in typical protogyny fishes. If the male is lost, the largest of the minor females start to court other females. It develops testes and other male sexual organs and mates with remaining females just as in regular protogyny. However, if a new male from a different school appears in the area, or more females develop into a dominant male, one or more males convert their sex back to the submissive female and continue to fulfill their reproductive evolutionary roles as females. Thus, instead of fighting and possibly dying, males convert back into the females and reproduce, a win-win situation from the evolutionary perspective.

In many fishes, multiple, bidirectional sex changes are more an exception than a rule, but not in some coral-dwelling gobies. In these fishes, bidirectional change of sex is frequent and not connected with social habits. Coral-dwelling gobies can change sex many times in either direction throughout their lifetime. It is not clear what drives these transitions of sex. The most likely explanation is, as these are small fish, living isolated in the coral reefs with limited opportunities for meeting and mating, that frequent change of sex is an evolutionary adjustment to enhance mating success. If two lonely males meet at the beautiful coral reef, it surely is advantageous from the evolutionary point of view if one could convert into a female. They can produce the brood rather than create a childless friendship.

Most men would probably prefer to be stranded on an isolated island with a woman and vice versa, women with men. Unfortunately, humans cannot switch sexes. If two men or two women were shipwrecked on a remote island, they could not start a family. In contrast, two male or two female gobies on an isolated coral reef could adjust their sexes and begin to reproduce.

Stereotypically, we often think of fish mating as a female releasing her eggs into the water, followed by males discharging their semen in the eggs' vicinity, without any physical contacts or romanticized rituals. Although this is true for many fishes, mating with bodily contacts and internal fertilization is not uncommon. Males have developed specialized organs, most often as derivative of one or two fins, analogous to the penis in mammals. In females, oviductal tubes, similar to oviducts in humans, connect the ovaries with the outside world. Unfertilized eggs are released into the environment through these tubes, or male sperm reach the eggs that remain in the vicinity of the ovary.

Most fish of the same species look the same to us, and if we look at the sardines in the tin, we will struggle to tell males and females apart. Yet, some fish males develop special features, akin to antlers in deer or colorful tails in the peacocks. In many salmon species, males in the mating season grow an elongated lower jaw with the hook at the end, called a kype. Kypes in salmon were observed and described already by Charles Darwin. He considered them a fighting tool for combats between breeding males. It is not clear if this is true, as male salmon are rarely seen fighting. Kype might be only an aesthetic appendage, just like a peacock tail or Armani jeans, trying to convince the female who would be a better mate. Sockeye salmon, spawning in Alaska, have another curious habit at the time of reproducing. When they return from the sea to their spawning grounds, the skin in males and females turn from its usual fishy silver-blue to bright red color. Chemically, this happens as carotenoids, natural pigments giving orange/red color to carrots, corn, and shrimps, accumulate under the salmon's skin. The brighter the red color, the healthier is the individual, thus more fit to reproduce and more successful in luring females into the mating.

Similarly, three-spined stickleback males develop bright red-orange breasts during the mating season. Sex differences are also present in many other fishes in the form of body size, shape, or size and color of fins. However, they are often not as prominent as in salmon.

In mammals, including us humans, courtship behavior is regularly an essential part preceding any sexual encounters. The human male invites a female partner out for a drink and chit-chat, while the male deer is bragging with his large antlers in front of the female. In fishes, courtship behavior is not very common. It is almost non-existing in fishes that live in large schools and release their sperm and eggs into the water. It is also not common in fishes with internal fertilization. It does occur, however, in some species at the time of mating, but it is usually not very elaborate or lengthy. Most often, the male would swim around a female, flashing his fins in front of the lady. He might raise and lower the fins, touch or stroke the lady with the fins, or even butt her with his head until she is ready to mate.

Some fishes, however, perform extravagant courtships. Seahorses have one of the most elaborate courtships in the underwater world. The romance in seahorses may last for several days, during which males and females swim side by side, change colors, and hold with their tails on the same seagrass. When they are ready to mate, they position themselves face to face. Male and female look almost like they are kissing each other, and they swim or rather drift upward toward the surface of the water. After several repetitions of this behavior, the female inserts her sexual organ, called an ovipositor, akin to the penis, into a male brooding pouch. The female seahorse then discharges her eggs inside the male. In seahorses, it is a male who is impregnated, becomes pregnant, and delivers babies.

Popular aquarium fish, Siamese fighting fish or betta, are, as the name suggests, very aggressive fish. Males and females often fight, but during the mating, male and female embrace in cuddly hugs, even though they have external fertilization. The cuddles stimulate the female to release her mature eggs, and males and females keep hugging until the gal releases all her eggs. When eggs are released, the bloke discharges the semen to fertilize the eggs. Then he collects fertilized eggs in his mouth to form the so-called bubble nest from his saliva and air. These bubble nests float close to the surface of the water. Male thereafter guards the nests against possible predators, including the mother of the eggs, as females often eat their own brood. Before mating, betta males often fight fiercely for the females, and females sometimes eavesdrop on the fights. Interestingly, eavesdropping females usually choose the winner male to mate with. In contrast, females who did not witness the battle and do not know who the champion is are not choosy and happily go with the loser.

Some fish have developed other courtship ceremonies. In several fish species, males build nests to impress females and lure them into the mating. Male Japanese pufferfish create beautiful sand sculptures consisting of numerous underwater circles. They can spend several days constructing them from the sand at the bottom of the sea with their fins. Drummer fish use their air bladders to produce low, croaking or drumming sounds. The music does not sound like a serenade to our ears. Yet, it is effective in charming female drummer fish and alluring them into mating. Male clownfish do not build sandcastles, but they tidy up and clean a bare rock near their anemone to impress the female.

After completion of mating, developing eggs and fish youngling face various fates depending on the species. In some, mostly in pelagic fishes (fishes that live in the open ocean waters), eggs are released into the sea, fertilized by passing by males. The eggs are then left to their fate in the ocean without any care from their parents. The majority of these eggs and younglings die, mainly as other fish eat them. But fish parents counteract this by releasing an enormous number of eggs. Hence, at least some develop into fully grown fish. In some species with internal fertilization, mothers release fertilized eggs just like chickens and other birds. These fishes are called oviparous from Latin—*ovum* meaning egg and *parous* meaning birth. Some other fishes are, like mammals, viviparous, meaning they give birth to live offspring that develop inside their reproductive organs. Other fishes are called ovoviviparous. In such fish, fertilized eggs remain inside their mothers' bodies, but fetuses remain inside the egg. They only receive protection from the mother while all the nourishments come from the egg, just as if the egg would develop outside the body.

In viviparous fishes that give birth to live babies, developing fetuses obtain food inside the mothers in various ways. In some fishes, mothers provide nutrients for the young through an organ similar to the placenta in mammals. In others, the fetuses consume the mothers' tissues, and sometimes they are fed by unfertilized eggs inside the female. In some, the strongest fetuses feed on less developed, weaker fetuses, what is called intrauterine (inside the womb) cannibalism. Probably the most interesting viviparous species are seahorses and their relatives, pipefish. In these fish, females deposit their unfertilized eggs inside the unique male organ called a brooding pouch. Inside the pouch, eggs are fertilized by male sperm and then develop inside the male until young seahorses are ready to face the outside world. The pouch is not only protecting developing younglings but also provides nourishment. Interestingly, similar genes that regulate changes in the uterus in pregnant mammalian females control the development of the seahorse pouch to provide for the babies.

After eggs are laid, or babies are born, they face different fates depending on the species. Although most fishes do not take care of their eggs or younglings, about one quarter of all fishes show some kind of parental responsibility. Parental care varies from simply hiding fertilized eggs in the sand, as trout and salmon do, to building elaborate constructions protecting eggs and younglings. Some fish even hide the babies in the mother's or father's mouth. Interestingly, in fishes, fathers more often take care of their young than mothers. However, most fathers do not go to such extremes as seahorses and pipefish, where fathers get pregnant and deliver babies. In clownfish, the father builds the nest near the anemone by clearing the bare rocky spot. After the female lays the eggs, the male protects the eggs from the predators and fans the water to provide more oxygen for hatching eggs. Once the eggs are hatched, usually after 6–10 days, younglings swim away without further parental care. However, some fishes build much more elaborate nests and care for their young for extended periods.

Cichlid fish from African Lake Malawi build nests from sand that look like craters and can be up to 3 m in diameter. Three-spined stickleback males, fish that live in various rivers in the northern hemisphere, build large nests at the riverbeds. They use sand, plants, and special glue-like substance from males' kidneys as a building material. Male firstly builds a sizeable covered building with walls and roof and then makes an entrance, a narrow tunnel that leads inside the nest. The male then begins to court the passing-by females by making unique zigzag swimming in front of his residence. If the female responds to his courtship, he leads her inside the love nest to lay her eggs. The male immediately fertilizes the eggs with his sperm and afterward chases away the female. He then takes care of the eggs by fanning the water near the entrance, thus providing more oxygen in the water inside the nest. He also chases away any males, predators, and also females that do not carry eggs. However, if another gravid female approaches, he would let her enter the nest, and she might deposit additional eggs. One nest could, therefore, contain eggs from various mothers but a single father.

Once eggs are hatched, the father takes care of the younglings for several days. He keeps them inside the nest, and if any of them escape, he gently grabs them by his mouth and spits them back into the nest. After a few days, when younglings are large enough, they leave the nest and wander away. Father then usually leaves the nest, although occasionally, he will repair it and use it for the new reproductive cycle.

Some other fishes such as marine catfishes and cichlids (tilapia and angelfish) do not build nests but protect eggs and younglings in their mouth. Some cichlid fishes collect the eggs in their mouth immediately after spawning, and younglings hatch inside the mouth. Other fishes leave the eggs in the water until they hatch, and only then parents, usually mothers, collect them inside their mouths. The infants are kept inside the mouths until they are big enough to survive on their own. This offers the younglings perfect protection, as possible predators could get to the younglings only by killing a parent.

For many people, some of the most bizarre marine creatures are anglerfish. These fish dwell very deep in the oceans, sometimes a mile or more below the sea surface, where no sunlight is present. They are curious creatures with large teeth. They are called anglerfish due to appendages on their heads, often illuminated by fluorescent lights. Fluorescent lights attract various other deep sea-dwelling creatures into the vicinity, where carnivorous anglerfishes could attack and devour them. But these shiny appendages and large, scary teeth are not the only curious thing about the anglerfishes. Naturalists were puzzled about their reproduction for a long time as they could identify only female individuals, with no males in sight.

Still, scientists knew they are not hermaphrodites. Genetic analyses have shown that offspring are not genetic clones of their mothers. Finally, the puzzle was solved, and scientists discovered that males also exist in anglerfish, but they are literally insignificant. In comparison to the females, males are extremely small, and couples develop an extraordinary and very close relationship. Tiny males are born with sharp teeth and excellent senses, similar to our noses and eyes. These senses help males locate the females in the dark ocean waters, where sunlight does not penetrate. Once the male finds the female, he grabs her with his sharp teeth, and gradually, his tissues fuse with the female's tissues. They merge their blood vessels, and the male becomes entirely dependent on the female. He gets all the nutrients and everything he needs for living from the female. He loses all the other organs such as the brain, eyes, and others, as they are no longer required. Males turn into some kind of a female's organ, with its sole role to provide sperm as needed. In this fish, we can say not only figuratively that males lose their brains from being in love.

Cold, wet fishes are not usually associated with love and romanticized relationship. However, in numerous fishes, males and females show courtship rituals, live together, and take care of their young. Interestingly, fathers are often better parents than mothers, but little is known what drives courtship, nest building, monogamy, and parental care in the fishes. Some similar hormones, as in mammals, including us humans, likely have some role in governing the social instincts in fish. Hormones oxytocin and vasopressin, connected with parental behavior in mammals, and possibly love and attachment in humans, do not exist in fishes. But their evolutionary predecessors, arginine vasotocin and isotocin, likely have some roles in regulating reproductive and social behaviors in fish. The hormone prolactin, responsible for milk production in human mothers and with a role in the feelings of attachment, regulates fanning behavior, often observed in fishes caring for hatching eggs.

Interestingly, the male sex hormone testosterone is in mammals connected with higher aggression. More caring mammalian fathers have lower testosterone levels. In contrast, fish fathers have higher testosterone levels during the child-caring periods. This difference might be due to the differences in paternal care. Fish fathers do not cuddle or play with their offspring as we do. Their primary role is the protection of eggs or younglings from predators. Higher testosterone levels perhaps increase aggression in fish fathers, thus making dads better protectors of their brood.

Love certainly does not exist in fishes, but some early versions of hormonal systems that operate in our bodies are already present in the fish. Fish have sex hormones similar to ours that oversee the reproduction. Fish also have hormones similar to oxytocin and vasopressin that probably have some roles in parental care in fish. These are perhaps the early ancestors of mechanisms that developed into maternal love and ultimately into romantic love through an evolutionary progression from aquatic fishes to terrestrial humans. Over the millions of years of evolution, seeking the best ways to ensure the continuation of the species, nature slowly devised the means of protecting the brood. Various mechanisms led to a more and more elaborate social life, which ultimately culminated in romantic love in our species.

# Chapter 12
# Octopuses: Lonely Cannibals

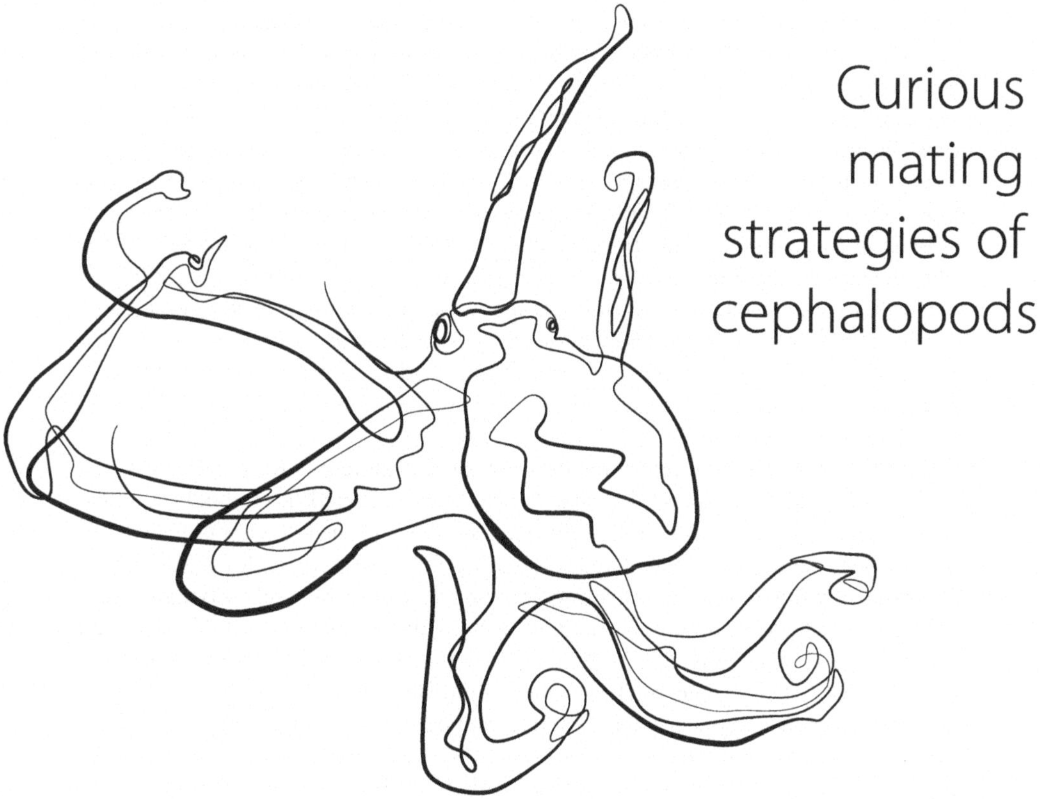


*Oh, I love hugging. I wish I was an octopus, so I could hug 10 people at a time!*

Drew Barrymore

Most of us remember Paul the Octopus, a famous cephalopod who accurately predicted winners of football matches at the 2010 World Cup. Paul correctly predicted the winner in 12 out of 14 games by swimming into the box with the winning nation's flag. Impressive performance, although it certainly happened by chance and not by psychic powers of Paul. Yet, the octopuses have some remarkable



G. Majdic, *Soul Mate Biology*, Fascinating Life Sciences,
https://doi.org/10.1007/978-3-030-67212-6_12

abilities and are most likely the most intelligent invertebrates. They have a sizeable nervous system, can perform some impressive tasks, and have some very peculiar habits.

The famous French naturalist, Georges Cuvier, wrote in the early nineteenth century after studying octopuses: "Here we have the body of the octopus which has for a parasite a worm so similar to the arm of the octopus that the illusion could not be greater…." However, Cuvier was led astray in his thinking by the strange biology of some octopuses. What he described as an illusion of the octopus tentacle created by the parasite's mimicry was indeed a tentacle that male octopuses use as a sexual organ. Peculiarly, during the mating with an amorous octopus lady, some male octopuses detach their sexual tentacle from the body. Although lady octopuses might be amorous, they are often very hungry and ready to devour even their loved ones.

To avoid such a grisly fate, to prevent losing their head due to being in love not only metaphorically but literally, some male octopuses developed the ability to detach their sexual organ. This trick lets them escape from the deadly embrace of their lover before they might be swallowed by love. This tentacle then stays inside the female reproductive organs until sperm are needed. When Cuvier found these foreign objects inside the female bodies, he understandably thought they must be a type of parasitic worm.

Georges Cuvier is famous for his descriptions of numerous fossils. He was one of the first scientists to argue that mass extinctions occurred throughout the eons. Although he opposed the idea of evolution, his contributions to the natural sciences are very important. He is considered a founding father of paleontology, a scientific discipline studying extinct living beings. In the late eighteenth and early nineteenth century, he spent some time in Normandy on the Atlantic Ocean coast. He observed various marine organisms, and he was the first to classify snails, various shellfish, and octopuses in the same phylum called mollusks. Despite considerable differences in appearance between snails, oysters, and octopuses, Cuvier noticed a similar body structure and correctly concluded that all these animals are related. Mollusks are the largest group of animals after arthropods (arthropods include insects, spiders, centipedes, and crabs), and Cuvier named them after their soft bodies. The name comes from the Latin word *mollis*, meaning soft, as all members of this phyla have soft bodies, even though many like snails and clams have hard outer shells.

Cephalopods, a class of animals that includes octopuses, squids, and nautiluses, are close relatives of shellfish and snails. Their name derives from Greek words for head (*kephale*) and feet (*podos*). They are a fascinating group of animals. As their name suggests, they only have limbs and heads, but no body, and all organs are stored inside their heads. This also includes their highly developed nervous system. We do not know whether octopuses can think. Yet, they do show some remarkable abilities, connected with their well-developed nervous system. They have a vast number of neurons. Because they have not developed an insulation layer called myelin coating like mammals have, some of their neurons are exceptionally large to enable faster propagation of the nerve signals. Giant squid neurons could thus measure up to 0.5 mm in diameter and could be seen by a naked eye, a feat not very common for a cell.

Another interesting feature of cephalopods is their eyes. If we stare in the octopus' eye, we would often have the feeling of looking into the human eye. This is because human and cephalopod eyes have very similar structures. Cephalopod eyes have iris, lens, interior chamber, and retina with optic nerves, just as human eyes do.

One notable difference is that cephalopods do not have a blind spot. In the human eye, optic nerves lie on the top of the light-sensing cells. A large nerve that sends signals from the eye to the brain must pass through light-sensitive cells at one spot, thus creating a blind spot in our eyes. In cephalopod eyes, optic nerves are under the layer of light-sensitive cells. The optic nerve, therefore, does not disturb the layer of these cells anywhere in the retina. Similarities between vertebrate and cephalopod eye are stunning, as we are evolutionary very distant. This similarity is, therefore, not due to the development from a common ancestor. Instead, it is a result of the so-called convergent evolution—an evolutionary process in which similar organs develop independently in different, evolutionary very distant, organisms.

Octopuses (and probably other cephalopods) are often a nuisance when studied either in nature or in the laboratory. Some octopuses can learn many tasks, but others don't. Scientists have the feeling that those not learning the jobs are not unable to learn but are just unwilling to cooperate with the scientists who are studying them. There are reports that octopuses have learned how to switch off the light by spraying the water with high pressure onto the light switch. Other octopuses learn how to block the drain in the aquarium, so the water overflows and helps them escape. There are even reliable reports about octopuses leaving their aquarium during the night and entering the neighboring aquarium. There they raid the fish for food and after the feast return to their own aquarium.

Octopuses can navigate simple mazes, solve certain puzzles, and open jars with screw-on lids. In the wild, octopuses are stacking rocks in front of their dens for protection or carry around an empty coconut shell in which they hide when danger looms.

Remarkable animals as they are, often triggering feelings of admiration and repulsion simultaneously when we observe them, they are very short-lived. Most of the octopuses live only a year or two. It seems a remarkable feat of evolution to develop such a complicated nervous system for such short-lived animals. Therefore, some argue that cephalopods do not possess special intellectual abilities but have an extensive nervous system just to control their bodies. Indeed, eight wriggly tentacles are undoubtedly hard to control. We should just imagine having 8 arms with perhaps 100 elbows in each of them, and we could understand the difficulty of controlling the cephalopod body.

In addition to the remarkable nervous system, some cephalopods also have curious mating habits. Most mollusks use external fertilization for the propagation of the species, but not the octopuses. Octopuses developed internal fertilization, and the male octopus has to deposit sperm right inside the female reproductive organs.

Most octopuses are solitary animals, always hungry, always ready to catch a new mouthwatering bite. They are not fussy about what is delicious. They would feast on a member of their species just as happily as on a fresh fish. Lustful female, in most octopuses larger than male and ready to mate, is often hungry. Mating could be, therefore, a hazardous proceeding for a male. For a hungry female octopus, a tasty snack in the form of a male, so close by during the mating, is hard to resist. Females thus frequently perform the kiss of death during the mating, using the male to fertilize her eggs and provide her with a snack after having sex. Female octopuses have been observed strangling their mates during the intercourse and then taking their dead bodies into the dens to feast on them after the sexual encounter. Surely, an interesting way to satisfy appetite after sex.

On the other hand, male octopuses are understandably not too keen on the idea of being eaten during or after sex. To avoid this, they developed mating strategies to eliminate the danger of being devoured by love.

Female octopuses have ovaries inside their heads (as all other organs) and have a special opening called siphon in their mantle. To perform successful mating, the male must come close to the female and insert his sexual organ called hectocotylus inside the siphon. Hectocotylus looks in most octopuses like a regular tentacle. When not used for mating, this organ performs the same function as the other seven tentacles. Yet, inside this specialized tentacle is a deep grove, a structure with similar function as mammalian vas deferens, a tube specialized for transporting sperm from the testis into the female reproductive system. Hectocotylus also has erectile tissue at the end of the tentacle, not unlike the erectile tissue in the human penis. This erectile tissue also serves a similar function, enabling erect hectocotylus to enter the female siphon.

During the mating, the male releases packet of sperm called spermatophores (like in insects) into the oviductal gland close to the ovary, where sperm are stored for some time. Females could release their eggs from the ovaries weeks or even months after the mating. When released eggs commute through the oviductal gland, they become fertilized by stored sperm. Males sometimes mate with females before females reach sexual maturity. Due to their short life span, males might be already dead when their sperm fertilize the eggs of their loved ones for the first time.

In general, male octopuses developed two mating strategies. In octopuses with shorter tentacles, mating is usually a close encounter. The male approaches the female from the back and mount her on the top of her head. After he grabs the female with seven tentacles, he inserts the eight, erect hectocotylus into her siphon. He then releases the spermatophores as quickly as possible and runs, or better swims, for his life. The mating lasts only for about 2 min in some small octopuses, while it could last up to 30 min in larger octopuses. During this time, the male tightly embraces the female, probably not as much as the display of affection but more likely to prevent her tentacles from strangling him.

Octopuses with longer tentacles developed different strategies. They perform mating at a distance. Male approaches the female yet stays at a safe distance and only extends his hectocotylus toward the female. With the tentacle's erect tip, he enters the siphon and deposits his spermatophores inside the female. Mating at a distance is much safer for a male. Not only because he is not too close to the potentially hungry female, but he also has seven remaining tentacles available to help him swim off quickly or to defend himself if the female becomes too interested. Long-tentacled octopuses could perform the mating by not even entering the den of the female. Thus, they are not exposing themselves to danger. They often mate with a female when she is busy foraging for food and therefore distracted.

Mating from a distance has potential benefits for females as well, as one female could mate with two males simultaneously. Siphon is a paired organ, each female having two, one on each side of her head. Therefore, two males could extend their hectocotyli into one siphon each, but not because octopuses are incredibly kinky in their sexual habits and like orgies. In general, mating with different males from the evolutionary standpoint improves reproductive success. Mating with two males increases the number of genetic combinations in the offspring of such females. Increased genetic diversity is always beneficial for the survival of the offspring and thus for the survival of the species.

Some male octopuses developed a distant mating strategy even further to be on the safe side and be able to escape ravenous females quickly. They willingly sacrifice part of their bodies, their hectocotylus. The detachment of the hectocotylus occurs in several octopuses, such as argonauts and blanket octopuses. In both species, there is a tremendous sexual dimorphism in their body size. In blanket octopuses, females can reach up to 2 m, while males are often the size of a walnut. The difference in argonauts is somewhat smaller, but females are still many times larger than males. Perhaps this massive sex difference is why male octopuses of these two species developed a strange mating strategy of sacrificing their male sexual organ. Just imagine a creature hundreds of times larger than you want to eat you! A detachment of the penis might be a small price to pay.

Immediately after the male places the hectocotylus inside the female siphon, he detaches his organ. He swims away, leaving the hectocotylus filled with spermatophores inside the female siphon. The hectocotylus seems to be alive for some time after detachment. It is wriggling its way autonomously deeper into the female to place itself inside the oviductal gland. The hectocotylus can remain inside the oviductal gland for some time, providing the sperm to ovulated eggs whenever the female needs them. As lady octopuses are not faithful lovers, they might mate with more than one male. Therefore, sometimes, more than one hectocotyli are present inside the female's oviductal gland. However, how does the male perform this detachment of his body part is not well understood.

Although octopuses are generally very good at regenerating parts of the body, mating in argonauts and blanket octopuses has dire consequences for the males. They often die shortly after the mating interestingly, not due to the damages from the detachment of hectocotyli. They probably just fulfilled their biological role, inseminating the female and propagating their genes into the next generation. Female octopuses, in general, share a similar fate. In some species, females die shortly after laying the eggs. In other species, the females care for the eggs, clean and protect them, but once the eggs hatch, mothers die and leave the hatchlings on their own. So, there is no parental care in octopuses, and younglings have to learn everything by themselves.

Yet, before mating, octopuses display a remarkable ability to regenerate body parts. Octopuses with less than eight tentacles are seldom observed in nature. A new tentacle quickly replaces any amputated limb, regrowing from the stump. Specialized sex tentacles, hectocotyli, regrow even faster than regular ones. They are also less often damaged, likely as males protect them during foraging by keeping them close to their heads. It is interesting why hectocotyli do not regrow after mating in argonauts and blanket octopuses, why animals mate only once. But these are just some of the many unanswered questions about the octopus' world.

Love comes in many forms and so is true also in octopuses. While, in general, mating is not a romantic affair with these eight-legged creatures, some octopuses defy the general rule. The larger Pacific striped octopus is a species that was firstly described in the 1970s. Their behavior is so untypical for octopuses that a biologist from Panama who initially observed them could not publish his finding in scientific journals as nobody believed him. About 40 years later, scientists from the USA confirmed his initial observations. It is now accepted that larger Pacific striped octopuses are an unusual exception in the wriggly world of tentacles.

Larger Pacific striped octopuses are very social animals, living seemingly happily in groups of up to 40 individuals, highly unusual for generally solitary octopuses. There are no reports about cannibalism, and even mating looks like a very romantic affair. Unlike in other known octopuses, where mating is dangerous for males, and thus males try to make it as short as possible and with as little bodily contact as possible, larger Pacific striped octopuses behave completely different. Males and females of this species seem to like intimacy. After some mating rituals that include changing colors of their skin, they swim close to each other. They position themselves facing beak to beak and embrace each other with their tentacles. They could remain intertwined in the hug for up to 2 days, performing mating. However, it is not clear whether a portion of this long period might also be needed for males and females to disentangle from the embrace. Sixteen wriggly, knotted tentacles might be a challenge to untangle.

Males and females do not only share hugs. They eat the food together while in the embrace and spend several days together in the same den. Partners are seemingly cohabitating without a male being threatened at all. Another interesting observation is that females of the larger Pacific striped octopuses mate and lay eggs multiple times. In most octopuses, females die shortly after laying their first clutch of eggs. However, larger Pacific striped females could mate multiple times and lay eggs many times over several months. Yet, we still know so little about these fascinating animals. This lack of knowledge suggests that other surprises might wait for us in other octopuses whose behavior has not yet been observed.

Octopuses' fame was, and still is, predominantly being an asocial, ruthless, cannibalistic group of animals. However, the recent descriptions of the larger Pacific striped octopus change that. These octopuses demonstrate that there is a place for bonding even in a group of animals that have one of the worst reputations for their social and intimate behavior. Behavior so ruthless that males in some species have to sacrifice their sexual organs to avoid being eaten by their mate. But males and females of the larger Pacific striped octopuses embrace themselves in hugs that resemble an intimacy in our species.

We do not know anything about these unique relationships yet. It would be fascinating to learn whether there are real feelings of bonding involved in these animals and whether there are any similarities between genetic and hormonal regulation of bonding in mammals and sexual relationships in the larger Pacific striped octopuses. It would be a stunning discovery if we would find out once in the future that perhaps not only eyes in cephalopods and humans developed similarly but also genetic mechanisms regulating bonding evolved through convergent evolution. But this is, at the moment, pure speculation. We will have to learn a good deal more about both the nervous system and the brain of the octopuses before we will understand their mating habits and perhaps their feelings.

Surely, what we can see in the larger Pacific striped octopus cannot be called love. Still, most of us would have a hard time not to see and feel that there is something like affection in the intimate embrace between male and female larger Pacific striped octopus. But what governs and regulates such behavior in one species of otherwise unsocial and ruthless animals is unknown. Yet it shows there is a time and place for socialization and attachment everywhere around our beautiful planet.

# Chapter 13
# Graceful Swan: A Symbol of Love

## Mating and attachment in birds

*"Great testicles for the bigness of its body, whence we may infer that it is a salacious bird,"* wrote Francis Willughby and John Ray about the common quail in the first-ever encyclopedia of ornithology in 1676. Willughby and Ray were correct in their reasoning, as quails are indeed promiscuous birds. Males are mating with many females and do not live in monogamous pairs. But this is unusual in the world of birds, as the majority of birds live as couples at least parts of the year, and they develop seemingly devoted relationships.

In a small Croatian village, a female stork named Malena greets her partner every spring. She has injured wings and cannot fly, so she remains in Europe through the winters. Yet, every spring, almost every year exactly on 24 March, her partner returns from Africa and joins her in the nest. One year, he was late, and while she was alone, the female succumbed to the advances from another male. She mated with him, although this new male later left her and built a nest nearby, together with another female. However, when the old husband returned, he somehow found out that the new neighbor messed with his sweetheart. He became enraged and destroyed the nest of another couple and forced them to leave the village.



G. Majdic, *Soul Mate Biology*, Fascinating Life Sciences,
https://doi.org/10.1007/978-3-030-67212-6_13

Similarly, a male stork was causing problems for the villagers in a small German town Glambeck some years ago. The male behaved violently. He attacked cars and houses, but most of his rage was directed toward another stork couple in the village. It turned out that this enraged stork was a disappointed suitor who was jealous of another male. He was so violent that he finally forced another male out of the nest and the stork became peaceful again once the lady stork accepted him as her partner.

Storks are not the only birds living in relationships. Two swans, swimming closely together, with necks curved to create a form of a heart, are often seen as a symbol of love and faithfulness. The pictures of swans forming hearts are not made by Photoshop. Swans really perform mating rituals by being close together and form a shape of the heart with their gracious, long, curved necks. Indeed, monogamy, the partnership between one male and one female, is common in swans, as it is in many other birds.

The birds are the group of animals that most frequently live as committed couples. While only around 5% of all mammals are monogamous, over 90% of birds live in pairs, one male and one female socializing and taking care of their brood in unison. However, these relationships are usually not long-lasting commitments. Although some birds like swans often stay together until death do them part, most birds do not mate for life but rather find a new partner every breeding season. And not only that, even though they live in a strictly monogamous relationship, most birds are not the beacon of fidelity. With the advent of genetic analyses, we obtained the tools to accurately examine the relationship between individual animals. These tools showed that birds are not as faithful as we previously thought.

In many birds, chicks in the same nests do not always belong to the same father, partner of the female. So, it seems that in birds, there is quite a sharp distinction between social and sexual monogamy. While the vast majority of birds live in social monogamy, they are not sexually monogamous. Both males and females have sex with different partners during one breeding season while sharing their household with one partner only. Nevertheless, males from monogamous families do not seem to be concerned about their partners' extramarital affairs. They happily take care of the fledglings even if they are not their genetic progeny. After all, even though they will not pass their genes into the next generations, they will ensure the survival of the species.

Some birds take their marital vows seriously and pair for life, with swans being one of them. All seven species of swans form long-lasting relationships that seem to be socially and mostly also sexually monogamous. One exception to this rule is Australian black swans. One out of seven eggs on average does not belong to the black swan lady's partner. This suggests that black swan mistresses are not immune to the improper advancements from lovely, handsome mister swan from the neighboring nest.

Swans sexually mature at 3–7 years of age, although they might start to choose the partner when they are only 2 years old. When a potential swan couple meets, they perform a very long courtship ritual. Male and female swim together, perform synchronous movements, face each other, look each other in the eye, and curve their necks. After 20 min to 1 h of what could be quite a loud ceremony even in mute swans (who are not mute), swans finally mate.

After the initial sex, couples stay together, often for the remaining of their lives. However, divorces do happen in swans as in humans. If there was a lousy breeding season for whatever reason and swan couples were not able to raise the cygnets, swan couples sometimes split apart. Still, the divorce rates are much lower than in humans. In swans that bred successfully before, the divorce rate is around 3%. In swans that never had a brood, the divorce rate is about 9%, all much less than in the humans. In the USA, for example, every second marriage ends in a divorce.

Swans will find a new partner if they widow, but the mourning periods are very different between males and females. Males will typically wait until the next breeding season. They will start to search for a new mate in the regular mating season during the first winter after the loved one's death. On the other hand, females often find a new partner already 2–3 weeks after the demise of their previous spouse. To be fair to swans, this is most likely not associated with the lack of mourning or insensitivity for their loss. Rather, it is from the necessity to care for the young. While a male swan can defend the nest and the cygnets on his own, females are more vulnerable. Therefore, they need protection from a new male partner to keep the cygnets alive.

Other known birds that pair for life are black vultures, bald eagles, Laysan albatrosses, macaws, whooping cranes, California condors, barn owls, several penguins, and finally, clowns of the sea, Atlantic puffins. All these birds form lifelong bonds between males and females. Most of them are socially and sexually monogamous, meaning that chicks do belong to the parents who care for them. They might even show signs of affection that we could interpret as romantic and cuddling. However, in reality, birds mate for life for practical reasons and are not searching for a romantic soulmate.

Staying with one partner year after year eliminates the time and effort of finding and choosing the new partner every breeding season. Mating in birds is often an elaborate display that consumes energy and time in males and females. Males have to show off their most beautiful plumage and complicated dance or sing loud and beautiful songs. All these activities consume a lot of energy and expose males to predators. For females, finding the right male takes time and energy, which could be otherwise spent on caring for the brood. Therefore, it makes perfect sense for birds to remain in the relationship over the years. So why are not more birds doing it? There is a downside to being sexually faithful all your life—a reduced number of new genetic combinations.

The same male and the same female will produce offspring with similar genetic make-up year after year. If both partners have the same bad genes, only these genes are carried into the next generations. Their progeny would consequently have lower chances of survival. If partners are changed every year, this dramatically increases the number of new genetic combinations and makes the species more likely to survive. So, there is always a trade-off between being faithful to your partner and spreading your genes around. Both options have benefits and downsides, and it is up to the individual species to decide which option is the best for them.

Couples in some of the birds living in long-lasting relationships live alone. Others like albatrosses, Atlantic puffins, penguins, and vultures live together in large colonies, comprised of faithful pairs. These colonies usually have specific, sometimes interesting, rules. Colonies of black vultures seem to be very conservative societies. Individual vultures in the colony are well aware of who belongs to whom regarding the relationship. If a male or female black vulture cheat on its partner, other group members attack and ferociously punish the cheating partner. They do not stone the cheating party as is a suggested punishment for adultery in the Old Testament. Still, sharp talons and hooked beaks are just as effective as stones. So, no infidelity when in the company of the black vultures, please!

Male and female albatrosses, Adelaide penguins, and Atlantic puffins, unlike many birds of prey and macaws, do not live in the same place throughout the year. Adelaide penguins make very long trips between winters and summers to migrate from their summer breeding grounds to the winter Antarctica shores. Winter shores are many miles from the actual shores due to massive ice sheets forming over the sea during the winter. Penguins can find food during the harsh south-pole winters only in the unfrozen sea. They have to move hundreds of miles from their breeding grounds during long Antarctica winters to find an open ocean. When on the move, Adelaide penguins do not travel in pairs, and throughout the winter, males and females do not keep in touch. But when they return to their breeding grounds in the spring, when penguin love is in the air, couples from previous seasons recognize each other almost instantly and start a renewed family life.

Atlantic puffins spend most of their time out in the sky above the open ocean. They only come to the rocky shores in Scandinavia, Greenland, and Canada in the spring for breeding. As with Adelaide penguins, even though male and female puffins were separated since the previous breeding season, old couples find each other and start new cozy life in their burrows. How these birds find each other after several months of separate life is still not completely understood. Adelaide penguins most likely recognize each other by their voices. Puffins might just meet in the same burrow year after year, so they only have to find the home and the partner comes with the house. Once puffins meet in the burrow, they warmly greet each other with cute tapping of their bills together, called billing.

The majority of other birds do form monogamous bonds, but usually just for one season. Once the chicks grow up and could be left on their own, couples split apart and find new partners when the new reproductive cycle approaches. Male birds are usually excellent fathers and providers for the family. They help build nests, collect food, and bring provisions to their partners and younglings. They protect the partner and the brood from any potential danger. In many birds, males even help females in incubating eggs.

Despite this devotion to each other, neither male nor female is resistant to an extramarital affair. Genetic analyses showed that some younglings in the nest often do not belong to the father in the nest. Strangely, sometimes chicks do not belong even to the mother! Some songbirds' females, not just cuckoos or cowbirds, famous for being sneaky with their eggs, slip their eggs into the other nests of the birds belonging to the same species. It is not understood why they do this, as these females are not like cuckoos, who do not even make their nests and have to rely on other birds to hatch their eggs. These birds are generally good mothers, but they still slip some eggs into the other females' nests. Although purely speculative, one theory suggests that by spreading their eggs around into neighbors' nests, they increase the chance that at least some of the eggs will hatch and some of their offspring will survive. Even if something happened to the mother and her brood in the home nest, some offspring carrying her genes would have the chance to survive, being reared by the foster parents.

Birds are not only a group of animals with the highest rate of monogamy. As males have to impress and find a single partner, many male birds have amazing features, which amaze love-seeking females. Colorful plumage, long feathers, and beautiful songs, all these are just tools to enchant females and convince them to say yes. They have the same role as bouquets of red roses, sparkling diamond rings, or flashy sports cars in humans. Like our mating rituals when we try various things to impress our loved ones, male birds sometimes go a great length to win the loved one's heart. Male frigatebirds have large red pouches under their beaks. When males try to impress the female, they fill this pouch with air to look like a giant red balloon. Males with the largest pouches usually win the hearts of the females. However, males compete with each other. Sometimes males pierce the pouches of other males, so pouches deflate, and such males are less attractive to the females.

Flamingos are social birds and like to stick together. This is also true during the mating season when they perform mating dances in large groups. They flip their heads back and forth, stretch their necks, and make tiny steps. After a while, a female chooses her favorite dancing male. They retreat to a more secluded place where they perform copulation. Albatrosses perform ostentatious ritual dances, using their beaks to stroke each other's feathers, and clap with them like Spanish flamenco dancers with the castanets. We already mentioned billing in Atlantic puffins, yet white-fronted parrots go even further. They bring beaks together and touch each other tongues, performing something akin to a French kiss. But this is not all. After a while, the male vomits into the female beak, and no, she does not find this disgusting. The tasty treat gets the female into the mood for further action.

One of the most beautiful courtship rituals is undoubtedly performed by some grebes. These water birds make something like a gracious ballet dance before mating. Male and female grebe begin their mating ritual by swimming together and making similar moves with their heads and legs. After a while, they rise and begin to walk on the surface of the water. They move so fast, with 14–20 steps per second, that they tread on the water. Walking on water is further aided by the short flapping of their wings. They can cover distances longer than 60 feet, looking like a female ballet dancer stepping on their toes. Once male and female are certain they found the soulmates, they dive together under the surface. They often reemerge on the surface with some grass in their beaks, perhaps giving each other bouquet or just showing each other they are ready to build a nest together.

Bald eagle courtship is a striking and breathtaking ceremony, requiring a lot of trust and devotion from both partners. When two majestic US national birds find each other, they soar high into the air. Then they lock their talons together and begin the cartwheel display, or, as it is sometimes called, a death spiral. The couple spirals and tumbles down toward the earth in virtual free fall, with talons firmly locked. Just before they hit the land, they release the claws and fly back into the sky. Sometimes, though, eagles are too brave and crash on the ground as they do not release the talons in time. A death spiral and near-death experience may seem a strange courtship ritual, but let's not forget it is all about finding the best-looking and the most stalwart partner for your progeny. So, tumbling and free falling through the air might just be a test of bravery, stamina, and strength. After all, human males, usually teenagers, often put themselves in dangerous situations trying to impress their peers or loved ones. We drive fast cars or take our loved ones on roller coaster rides. Perhaps our courtship rituals, if looked from a distance, are not so different from bald eagles.

Bowerbirds, medium-sized birds with a rather everyday look, have one of the most interesting courtship rituals. Males do not parade in front of the female by colorful plumage, long feathers, or beautiful singing. Instead, they built elaborate structures, not for nests but just to impress females. Different bowerbirds use different masterplans for their buildings. Some build towers by placing small sticks around young saplings, and the others build two walls of vertically placed sticks.

Males also decorate these structures with various objects such as snail shells, flowers, feathers, stones, berries, and even dead beetles. In modern days, bowerbirds also use modern materials to decorate their buildings, such as coins, pieces of plastics, plastic bottle lids, nails, pieces of glass, and similar. Some males show a preference for colors, hoping a particular color will impress the female. Sometimes, they paint the structure with some berries. Today, they often use human litter left in the woods. They sometimes collect only blue pieces of plastic waste such as bottle caps and drinking straws and arrange them intricately around their constructions. However, males competing for the same females also play dirty games. They destroy each other's structures and steal their trinkets until they find a way to the female heart.

No bowerbird males go to a greater height to win the female hearts than MacGregor's bowerbirds. Unlike other bowerbirds who build their bowers in a few days, MacGregor's bowerbirds build elaborate towers, sometimes more than 1 m high. The construction of such towers from small twigs takes a long time, sometimes more than a year, but it is worth the effort and time. The male with the highest and the most decorated tower wins the hearts of the females. As bowerbirds are not monogamous as most other birds are, there is especially fierce competition between males. They sometimes use dirty tricks, ruining each other's bowers.

It is not only the height of the bower that matters. Males decorate the bowers with moss, lichens, flowers, and nuts. They even kill beetles only to decorate the bower and hang caterpillar feces from the twigs, everything to impress the madam. Once the female notices the bower, the male's struggle is not yet over. He has to impress the lady with his ritual dance. He shows off his colors and sings songs, sometimes most unusual such as imitating a dog barking or even human chatter. He plays hide and seek until finally the impressed female succumbs and agrees to mate. After a few short seconds, the sex is over. The female takes off, build the nest, and care for the chicks alone, while the male tries to lure more females into the mating.

All this effort of building, decorating, and dancing for only a few seconds of sexual pleasure! Of course, the ultimate prize is not a few seconds of sex. Victorious males transfer their genes into the next generation, and that is the highest prize in the constant evolutionary struggle.

For many ducks, including mallards, mating is far from being a romantic affair. Unlike in many other birds, mallards and other duck males do not waste time with an elaborate courtship ceremony. They simply approach the female, often hidden in a group of friends, and they literally rape the female. Very often, they perform gang rape, multiple males mating with the abused female repeatedly. Although the female cannot resist this gruesome violence, she can partially defend herself by choosing when and with whom she will have ducklings. Ducks are a rare bird species that possess a proper penis, but females have few tricks up their sleeves. Their wombs are very convoluted, often with several blind tubes that do not lead to the ovary and eggs. It seems that female ducks can choose where to direct the penis of a raping male. Female ducks must succumb to the gang rape, but they can prevent being impregnated by males they do not like and choose the father of their brood.

Most birds are socially monogamous, but some found that having more than one partner could also offer advantages. Dunnock or hedge sparrows adopt variations of social life, depending on the circumstances. Sometimes they live in monogamous pairs. Some live in groups of two males and two females, with both females copulating by either male. Sometimes, one male lives with two females, and often, one female is courting two males. In the last case, both males will usually take care of the female and her brood, as they do not know who the younglings' genetic father is. However, males sharing the female are aware of each other and often fiercely compete who will father the chicks.

As copulation in dunnocks lasts only one-tenth of the second (but on the other hand, they can copulate up to 100 times in 1 day), it is difficult for males to prevent sexual intercourse. Instead, after the female copulates with one partner, the other partner approaches her from the back. He pecks with his beak inside the cloaca, stimulating her to eject the sperm of the previous male.

During evolution, dunnocks also acquired some bodily changes that help them spread their genes more efficiently. A general rule of thumb in the animal world is that large testes imply females' promiscuous behavior. Indeed, the dunnocks' testes represent about 3.5% of their body weight, which is incredibly large. If we had the same ratio, both our testes would weigh between 2 and 3 kg, and each testicle would be the size of a large water or coke bottle! A larger testis means more sperm in the ejaculate. This is one way to ensure that your sperm will have a better chance than your competitors once inside the female reproductive tract. Only by chance, with more sperm there are more opportunities that one of yours will reach the egg.

Another adaptation common in birds with fierce competition between males is the length of the penis. The longer the penis, closer to the ovary and ovulated eggs sperm can be deposited. Hence, they have a better chance of being the first to reach the eggs. In ducks, where males often gang rape the female, the penis' length is especially important. Some ducks have developed remarkably long penises, up to 17 in. in the Argentinian lake duck. Another trick males developed to ensure the success of their sperm with promiscuous females is to monopolize the female. They do not enter into a single partner relationship but just prolong the copulation. Some birds, such as aquatic warbler and vasa parrot, have intercourse that lasts up to 30 min. Not because the couple is enjoying slow and tender sex. In that way, the male prevents other males from copulating with his partner. Male can hope that in 30 min sperm will reach the eggs inside the oviduct and perform their role before the female finds another sexual partner.

When we are in love, we often sing. Serenading under the window of the loved one is often seen as a symbol of love. From medieval bards to contemporary rock stars, musicians have always been popular with the opposite sex. But not only in our species. In many birds, females fall for the best singer, the male who produces the most beautiful song.

Songbirds are a large group of birds that uses songs to communicate. Although most birds emit some simple, monosyllabic sounds, these are named calls, not songs. Songs, on the other hand, are a complex compendium of voices produced by a single bird that follows specific patterns. Singing is mostly used for two purposes—to mark the territory and warn the potential intruders and attract females, potential mates.

Birds have a unique brain area that regulates the singing. This area is sensitive to sex hormones testosterone and estradiol. The bird song system is modulated by both male sexual hormone testosterone and female sex hormone estradiol. Although testosterone can modulate the singing in some birds, it is not completely clear what testosterone's real role is. In Europe and North America, only males usually sing, most likely under the influence of testosterone. However, in numerous birds, especially in tropics and in Australia, males and females sing, regardless of differences in sex hormones in their blood. Birds are highly seasonal, and when not in the breeding season, testes regress and produce very little testosterone. Yet, many male birds begin to sing at the very onset of the breeding season, before the testes fully kick in and start to produce massive amounts of testosterone. So, it seems that male sex hormone testosterone has different roles in the regulation of singing in different birds. Its roles and importance for singing likely differ from species to species.

Birds produce songs in a special organ called the syrinx, which is somewhat similar to vocal cords. While vocal cords are located at the top of the windpipe, syrinx is at the bottom of the trachea. More importantly, birds can regulate the left and the right syrinx independently, something we cannot do with our paired vocal cords. Singing birds can thus produce two different tones simultaneously, making them the envy of every human vocalist. Many singing males are true troubadours. They compete with each other who can sing a more beautiful song and attract the female. Females are listening carefully and evaluating individual songs. The best singer is awarded by being chosen to be a partner of the female.

Singing is not something that all birds are born with. Although birds are born with the brain circuits that regulate the singing, and some birds sing automatically, most of them have to learn how to sing properly from their parents or peers. They start learning while they are still in the nest, and this is the critical period. Birds that do not learn how to sing in these early days will never sing properly. Birds that are cross-fostered by parents from different species may sometimes sing different songs, not typical ones for their species.

Some birds could adapt their songs throughout their lives and include local sounds, not coming from the birds, into their singing repertoire. Catbirds, thrashers, and mockingbirds often include in their vocal expressions sounds from their environment like frog croaking, cats' meowing, and even car honking. They likely include these sounds to protect themselves and not to impress the females with a frog croak or loud car honk. Singing, although essential to find a mate, also exposes birds to potential predators. If they use sounds commonly heard in their environment, this might offer some degree of protection as the predator might not recognize such voice as a sound of the potential prey.

Birds also have dialects. Birds from the same species that live in different areas might have quite different dialects. They might not even understand each other if they meet, just like Scottish Highlander and African American from Alabama might have a problem communicating, even though they both speak English. But ultimately, regardless of dialect or car honks inserted in the song, the best singer wins, just like on "show your talent" TV show.

Birds display a remarkable variety of behaviors connected with sexual reproduction. If we want, we can see many similarities with human behavior. Although without priests or registrars, some birds mate for life. If things do not turn out as they should, they might also divorce. Some birds mourn their partners passing away but will nevertheless find a new partner and start a new relationship. Although most birds live in socially monogamous pairs, not dissimilar to humans, they often cheat on their partners, probably more often than we do. Some birds do not pair with their mates but remain single, and some mothers have to endure life as single mothers and bring up their fledglings alone. And finally, some birds live in a special relationship, sometimes made up of more than one male and female, just like some people live in communes.

But is there love in the birds' world? It depends on how we define love. Tumbling in high speed down from the sky toward the ground with the talons firmly locked undoubtedly requires a lot of trust and devotion, essential parts of any long-term relationship. Staying together year after year and recognizing each other immediately after weeks of gruesome trekking through Antarctica's unhospitable planes certainly seems like love and dedication.

The birds almost certainly do not dream of a prince on a white horse and happily ever after. Still, they show signs of sincere attachment and devotion to their partners. They most likely do not fall in love like we do, being almost crazy at the thoughts of the beloved person. Still, their long-term relationships are not unlike happy, long-lasting marriages in our species. So, we can probably learn a lot from birds. We parted on an evolutionary tree long ago, before the dinosaurs' age, and have developed in our separate ways for more than 200 million years. Yet some workings of the brain are remarkably similar. Birds do not have hormones oxytocin and vasopressin that likely have crucial roles in the attachment and bonding in mammals, including humans, but other hormones must play similar roles in birds' brains.

Birds have two similar hormones, mesotocin and vasotocin. These two hormones evolved alongside oxytocin and vasopressin from the common predecessor that existed in our ancestors already around 600 million years ago. It seems that these two hormones indeed play a role in the regulation of attachment and bonding behavior also in birds. Remarkably, even though we have evolved in our separate ways for more than 200 million years, both birds and at least some mammals have developed remarkably similar behavioral mechanisms to ensure the survival of the species. Love might be a paramount sensation that only emerged in us, conscious beings. But let's not forget that some birds like crows, ravens, and some parrots are remarkably smart. They have very large brains for their body size. They can solve complicated tasks and use tools, and some studies suggest they possess something similar to human consciousness. So perhaps, in line with the brain's development in birds, feelings of attachment, devotion, and bonding are also developing. If not yet present, they perhaps will grow into something similar to love in the future eons.

# Chapter 14
# Gosling Following Their Mothers

*The truth about an animal is far more exciting and altogether more beautiful than all the myths woven about it.*

Konrad Lorenz

You might stumble upon an extraordinary sight if you would take a stroll around northern Austrian town Altenberg around 1930. A full-grown man walking around in a crouching position and quacking loudly, followed by a group of ducklings. You would be forgiven you thought you had accidentally met a village fool. Yet, you would have the privilege to meet an exceptional scientist who was later awarded a Nobel Prize for his groundbreaking work on animal behavior. Today, he is considered the father of science about animal behavior. This person would be Konrad



G. Majdic, *Soul Mate Biology*, Fascinating Life Sciences,
https://doi.org/10.1007/978-3-030-67212-6_14

Lorenz, a famous Austrian scientist, one of the first scholars to study animal behavior in depth. He had a keen interest in animals from an early age, and he kept many different pets as a child. Although he studied medicine to fulfill his father's wish, he began to research animal behavior soon after graduating from the School of Medicine at Vienna University. During his studies and later during his early work in Vienna, he spent his summers at a family mansion in the northern Austrian town Altenberg. In Altenberg, he often walked around with a gaggle of goslings following him close on his heels.

In his experiments with geese and ducks, Konrad Lorenz observed that ducklings and goslings after hatching follow anybody willing to act as their parent, even if they are as different from them as humans are to duckling or gosling. This attachment of newly hatched birds is now recognized as imprinting, instinctive attachment between many newborn animals and anybody willing to act as their caregiver. Such imprinting is mainly present in birds, although with specific differences between species. Already Konrad Lorenz observed that goslings would follow him as long as he is present immediately after hatching. However, ducklings followed him only if he pretended to be a duck by squatting and emitting sounds similar to quacking.

Maternal love, or parental behavior, is essential in almost all birds and mammals. In the vast majority of animals from these two groups, newborn animals, like human newborns, are helpless and would not survive without somebody caring for them. For the survival of the individual (parent), this is somewhat paradoxical, as sharing the resources and energy with other individuals (children) diminishes the chances of an individual's survival. However, in birds and mammals, caring for the young is necessary for the survival of the species. Therefore, evolution invented specific ways to trick various animals and us into altruistic parental behavior. One exception from these is cuckoos and cowbirds, who developed a very innovative way to provide care for the young without investing their time and resources. As we all know, cuckoo and cowbird females lay their eggs inside the nests of other birds. Most birds are, in general, excellent parents. Therefore, birds unknowingly incubating cuckoo's eggs will care for the chicks that will hatch from the cuckoo's eggs and for their brood. This sneaking of eggs is called brood parasitism.

In many cases of brood parasitism, eggs of the birds that slip their eggs into the other birds' nests are similar to the hosts' eggs. In such cases, unknowingly adopting parents might have trouble distinguishing foreign eggs from their own. But this is not always the case. Sometimes, the parasitic chicks are larger than the host's and often throw some of the host female's brood out of the nest. So why do females tolerate such behavior and care for a foreign chick? At least in one species of cuckoos and one species of cowbirds, mothers of the eggs slipped into alien nests behave like proper mafia thugs. They linger around the nests of their hosts. If the host female destroys the foreign eggs, the parasitic mother, in return, annihilates the nest and eggs of the expectant mothers. Therefore, host mothers are better off accepting alien eggs. If they do so, they will at least keep their nest and their eggs intact. Hosts that have to deal with such mafia thugs are better off if they leave the parasitic eggs intact and care for the parasitic chicks. In that case, more of their chicks will survive than if they destroy the intruder egg and face revenge from the parasitic brute mother.

From the individual point of view, brood parasitism was a genius evolutionary development. It provides for the care of the younglings and frees the parents to spend their time taking care of their survival. One might ask why hasn't such behavior developed more often if it is so handy? The answer is straightforward. If too many birds would develop such parasitic practice, we would soon run out of affectionate moms and dads to take care of the young. Such evolutionary development would eventually become counterproductive. Therefore, in the majority of birds and mammals, parents take good care of their offspring. As a general rule, mothers always have a strong maternal instinct. Nevertheless, in many animals, especially in birds, but also in some mammals and fishes, males also show devotion to their progeny and help females rearing their young.

In most mammals and birds, parental instinct is not something that mothers and fathers are born with but develops when animals or humans are preparing to become parents. Hormonal changes during pregnancy influence the brain. Pregnancy triggers numerous changes in the activity of genes, sensitivity to hormones, nerve cells' activity, and even structural transformation of the brain. Brain cells change their shape and size, and they form new connections between neurons. New brain cells are born in some parts of the brain during pregnancy, with a special aim to promote parental love. The number of cells is reduced in other parts of the brain with the same purpose. Clearly, during pregnancy, the brain is remodeling itself. Undoubtedly, it is challenging to study due to the brain's complexity, which newly born or disappearing cells are directly connected with the development of the instinct for parental behavior. Nevertheless, it is now clear that structural changes do occur in the brain during pregnancy and are essential for proper parental instinct.

The main area of paramount importance for making someone a good parent is a group of cells called the medial preoptic area in the hypothalamus, the brain's autonomous center. The medial preoptic area is likely the primary regulator of the parental instinct. It induces and regulates proper maternal behavior toward the offspring. Cells in the medial preoptic area are highly responsive to different hormones present in the female body during pregnancy, and these hormones prepare the medial preoptic area to develop a parental instinct.

Cells in the medial preoptic area have extensive connections to other parts of the brain. Especially important seems to be a connection to the reward system, a brain circuit giving us a pleasing feeling when we do something good that is also important for our survival. The central molecule regulating the reward system is dopamine. This small chemical is released in the brain when we eat some yummy food, have sex, or do something else that makes us feel good.

Medial preoptic area, part of the brain that is important for parental behavior, usually does not have considerable influence over the reward system. But, when a pregnant female prepares for motherhood, the reward system becomes responsive to the signals from the medial preoptic area. These signals trigger the sensitivity of the reward system to the clues from the babies. When a mother is nursing the baby, close contact stimulates the reward system, triggering a release of dopamine and consequently a "feel good" response. This ensures that human or animal mothers want to spend time with pups or children and care for them, as they experience proximity with a child as a rewarding event. The reward system is thus safeguarding the survival of the offspring.

Another vital area in the brain connected with parenthood is the amygdala. The amygdala regulates our emotions, especially fear. Entering new, unknown places, seeing new objects somewhere they should not be, or getting in close contact with strangers activates the amygdala. The amygdala sends warning signals throughout our brain to make us more careful, prepare us to defend ourselves, or force us to run away if danger ensues. Around childbirth, in preparation for motherhood, the amygdala activity changes, and this has several effects. It prevents aversive response toward the newborn baby, what would be our natural response toward strangers. Alterations in amygdala activity also increase mothers' vigilance and preparedness to protect the baby. The amygdala's altered activity also regulates emotional reactions to the baby, making interaction with the child a more positive, passionate experience.

As any women know, profound hormonal changes occur during pregnancy, influencing the whole body. Throughout the pregnancy, high levels of female sex hormone progesterone are in women's blood, and estrogen levels rise steadily throughout the pregnancy. Toward the end of the pregnancy, the pituitary gland starts to secrete hormone prolactin, which stimulates the breast to develop functional cells that will eventually produce milk. During labor, the pituitary releases the hormone oxytocin, which helps with childbirth. Immediately after birth, progesterone levels fall rapidly, as it is produced by the placenta, which is discharged from the womb shortly after birth. Prolactin levels remain high, as this hormone is made in the pituitary and is necessary for milk production. All these hormones have a strong influence on the brain. Progesterone, estradiol, and prolactin prepare the brain to become responsive to the infant, forcing mothers to take care of the offspring after birth. After birth, a fall of progesterone levels is likely one of the triggers for mothers to start caring for their offspring. However, progesterone, estradiol, and prolactin do not act alone. Their actions converge around modulating the release of another hormone, or better neurotransmitter—oxytocin.

The primary hormone responsible for proper maternal behavior in mammals seems to be oxytocin. Oxytocin is a small protein molecule produced in the brain and secreted through the pituitary gland into the blood. A specific group of cells in the hypothalamus has a principal role in the production of oxytocin. These cells are unusual as they do not release oxytocin into the blood vessels in their vicinity. Instead, they send oxytocin on a trip inside their long, tiny protrusions called axons. These axons connect the hypothalamus with the pituitary gland, a principal endocrine gland. From the endings of these protrusions in the pituitary gland, oxytocin is finally released into the bloodstream and then transported around the body.

Many of us have already met with the hormone oxytocin. One of the essential functions of this hormone is during childbirth. Oxytocin is responsible for the contractions of the womb, what is needed for the newborn to be pushed out of the uterus. Due to this function, the English scientist Henry H. Dale initially recognized oxytocin early in the twentieth century. He found that extract from the oxen pituitary gland triggered the womb's contraction in pregnant cats. The substance was later named oxytocin from old Greek words *oxus* meaning sharp or quick and *tokos* meaning childbirth. Together, these words give a new term *oxutokia*—quick, swift delivery or labor.

According to some reports, extracts from animals' pituitary glands were used already in 1911 in humans to strengthen the womb's contraction during childbirth. Oxytocin (not derived from the animal pituitaries but produced in pharmaceutical factories) is still used with the same purpose today. It helps women deliver the baby by strengthening contractions of the womb.

The oxytocin role is not finished after the labor. For several months after childbirth, the pituitary gland sustains oxytocin secretion as it induces contractions of cells in the mammary glands. These contractions are needed to push the milk toward the nipple, where it is available for suckling by the baby. Every time an infant feeds on her mother's breasts, suckling on the nipples sends a neuronal signal to the brain to release oxytocin. In return, oxytocin triggers the contraction of smooth muscle cells around the alveoli, small sacks filled with milk, and this releases the milk. Though suckling and caressing the nipple are potent triggers for oxytocin release, psychological stimuli could also induce the discharge of oxytocin. The crying of a baby can be such a stimulus. All mothers know that sometimes the milk starts flowing from their breasts after hearing the baby's cry, even if a baby is in a different room. Sometimes, oxytocin is released in the absence of the baby. Just thinking of the baby could trigger hormonal response, and milk starts to flow even if the baby is not by the mother.

Specialized cells in the brain produce oxytocin. Part of the oxytocin is secreted through the pituitary into the bloodstream to affect the uterus during labor or mammary cells during breastfeeding/nursing. However, cells that respond to oxytocin are not only in the womb and in mammary glands. Such cells are also in the brain where oxytocin induces feelings of attachment. Oxytocin seems to be the hormone responsible for creating the bond between the mother and newborn offspring. This role of oxytocin was shown for the first time in 1979 when scientists injected oxytocin into the brain of virgin rats that usually do not display assertive maternal behavior. However, just one injection of oxytocin activated maternal instincts. Female rats began to build nests and took care of unfamiliar pups, licking them and carrying them into the nest.

Numerous studies have shown the importance of oxytocin for maternal behavior in various animals. Injecting oxytocin into the brain induces maternal behavior, and blocking the oxytocin action in the brain prevents attachment between mothers and offspring in many mammals. Mothers without the oxytocin will not care for their younglings, at least not as well as usually mothers would do. Many animal females do not show proper parental behavior unless they have delivered a baby, but giving them oxytocin can change this.

Although mostly studied in laboratory rats and mice, it is likely true that oxytocin can trigger maternal behavior in females in most mammals. In sheep, ewes are very fussy when it comes to caring for the lambs. Although lambs look all the same to us, ewes are excellent mothers. They unmistakably recognize their lambs (probably through pheromones) and do not nurse the lambs that are not their offspring.

Yet, if mother ewe receives oxytocin, it accepts unfamiliar lamb, cares for it, and nurses it. Oxytocin makes ewes better mothers that are willing to foster other lambs.

Oxytocin, produced in the hypothalamus, seems to be the molecule connecting different parts of the brain that have essential roles in regulating parental behavior. Studies in animals taught us that oxytocin acts in the medial preoptic area where it triggers the motherly instinct. Through the preoptic area, oxytocin affects the reward system and increases the secretion of dopamine. The increased flow of dopamine makes contact with a baby a rewarding experience. Oxytocin also changes the amygdala activity, preventing aversive response to the baby and making the mother more vigilant in protecting her offspring.

In addition to hormones, experience has essential roles in the expression of maternal care in mammals. In rats, continuous exposure to pups eventually triggers maternal behavior even in the absence of hormonal stimulation that occurs during pregnancy. However, maternal care in such rats is never as good as in mothers. Experience seems to be especially crucial in primates. When rhesus monkey females are exposed to infants, females that were never mothers do not care about the infants. However, monkeys that were mums in the past show strong affection for babies even if they are not mothers now. This suggests that hormones like oxytocin are probably essential for the initial appearance of maternal care in primiparous monkey mothers. However, once females experienced motherhood, maternal behavior remains embedded in their brains, and they do not need hormonal stimulation any longer to take care of the infants.

Although we cannot do extensive experiments in humans, there is some evidence that oxytocin also plays a vital role in human parental behavior. Oxytocin levels are high in human parents, both mothers and fathers, after birth. Experiments with intranasal treatment with oxytocin showed an increase in bonding with children in mothers and fathers after receiving oxytocin. Both mothers and fathers with higher levels of oxytocin spend more time cuddling and caring for their children. Imaging studies with humans showed that similar areas as in rodents' brains are active in human parents in the presence of children, such as the reward system. However, our conscious brain, the cerebral cortex, is also involved when we play with kids.

In many mammals, hormones during later stages of pregnancy induce the onset of maternal instinct, but this seems to be different in humans. Women appear to be, in general, more maternal than females in most other mammals. Handling of children or just looking at children's photographs increases the release of oxytocin in women regardless of whether they are mothers or not. This suggests that maternal behavior in women is partially embedded into our brains and not triggered exclusively by hormones produced during pregnancy. Adoptive parents, siblings, grandparents, and other relatives often grow attached to children and display similar affection as biological mothers. This further confirms that pregnancy hormones are not necessary to trigger maternal love in humans.

However, estrogens, progesterone, and prolactin, hormones flooding the female body during late pregnancy, modulate maternal behavior in humans as well. All three hormones stimulate and increase maternal behavior and attachment to the child. Like in animals, they act most likely by modulating the release of oxytocin and by modulating the sensitivity of various brain cells to the oxytocin. Progesterone, estradiol, and prolactin facilitate and increase parental instinct, but they are not essential to induce the maternal instinct in humans. As we are conscious beings, psychological factors also have the power to trigger the feelings of motherly love. However, even psychological factors work through the release of hormones such as oxytocin.

Only about 5–10% of all mammalian fathers take care of the offspring. This percentage is strikingly low compared to birds, where 80–90% of all species developed parental behavior also in fathers. In mammals, paternal behavior is most often present in primates, including us, humans, rodents, and dogs. Although males are not exposed to estrogens, progesterone, or prolactin during pregnancy, the male brain in species with paternal care also changes in anticipation of the younglings. New cells are born in certain parts of the brain, and hormonal changes occur even in the male brain. Like in women, secretion of oxytocin increases in response to children, possibly through the neural stimuli such as seeing and hearing the baby.

Interestingly, prolactin also increases in fathers, even before the babies are born, suggesting a psychological influence on the secretion of this hormone from the male pituitary. Male sex hormones appear to influence paternal behavior, too, but in the opposite direction than female sex hormones, estrogens and progesterone. Testosterone restrains paternal instinct in many mammals and humans. Human fathers with higher testosterone levels in their blood are usually less responsive to children and spend less time in close contact with children.

The role of testosterone in fatherly love was confirmed in rodents, where castration and consequent elimination of testosterone from the system increases parental behavior in males. In contrast, the addition of testosterone to both males and females diminishes parental instinct in both sexes. Gerbils and marmoset monkeys show a great variety of paternal behavior between males. Some gerbil's and marmoset's males display strong affection and care for the infants. In contrast, other males are not interested or are even hostile toward babies. Measurements of hormones in the blood of such males showed that parental males have much higher levels of prolactin and lower levels of testosterone compared to males that do not exhibit parental affection, suggesting the direct role of these two hormones in how males behave toward pups.

In many mammals, males are not only not helping females rearing their young but could present a danger for the infants. Usually, not their offspring, but males often attack and kill pups found together with the female if they are not his progeny. Such killings look like ruthless behavior, yet they are often present in the animal world. They could be explained by the evolutionary drive in each male to spread his genes.

Lionesses are ready to mate only every 2 years after their cubs grow to be independent. If a new lion takes over the pride where lionesses have young cubs, he would have to wait for a year or even two before the females will be ready to mate with him. But if he kills the cubs, all mothers will enter the heat soon, and he will be able to sire his progeny. So, from the male lion's point of view, it is only natural to perform infanticide. With it, the lion eliminates the genes of his competitors and gives himself a chance to become a father soon and transfer his genes into the next generation.

Such infanticidal behavior is present in many animals and is thought to be a purely evolutionary drive. As we already talked about earlier, in some animals, males even do not have to kill pups themselves. In some rodents and even in few primates such as gelada baboons, pheromones secreted by males could terminate the pregnancy before pups are born.

Infanticide is also happening in our closest relatives, chimpanzees. There is no clear answer to why chimpanzees do it. The most plausible hypothesis is that the reasons are similar to those of lions and some other mammals—eliminating the competitors' genes. In chimps, there might also be another reason—climbing up the social ladder in the community. More matings often increase social rank, so more females ready for mating could be an opportunity to climb up the ladder. However, chimpanzee mothers are not only the great caretakers of their children. They also know how to protect babies from strong and violent males. Lower-ranking males usually perform infanticide in chimpanzees as they try to climb up the corporate ladder. Chimp moms, however, have learned that befriending strong males with a high rank in their family helps to protect their babies, even when these males are not fathers of their babies. Therefore, chimpanzee mothers diplomatically befriend influential males to protect their infants from young usurpers.

When we think of parents taking care of the offspring, we usually think of mammals and birds, but not about other animals. However, not all other animals leave their young to their fate. As we learned in the cartoon Finding Nemo, parental care is indeed present in some fishes, and quite a few fishes display paternal behavior. Probably the best fathers in the fish world are seahorses in which the roles of the sexes are, in a way, reversed. In seahorses, females produce the eggs, but the males store them in a special brooding pouch where they are fertilized by males' sperm. Fertilized eggs develop and hatch inside the male's brooding pouch and are released only when hatchlings are fully developed.

Although paternal care is not so extensive in other fishes, males help protect the offspring in several fishes. Clownfish males extensively tidy up a nest. They meticulously clean up the anemone where clownfish live, preparing it for the brood. After the female produces the eggs, the male takes over. He cares for and cleans the eggs regularly until they hatch. After hatching, he swims around the brood, incessantly fan with his fins, disturbing the water and providing the younglings with oxygen-rich water.

In many others, especially invertebrate animals, parents do not take care of their young. In invertebrates, parental behavior is rather an exception than a rule. Nonetheless, parental behavior is also present in some amphibia, insects, and octopuses.

Both parents take care of the young in strawberry poison-dart frogs that live in tropical rainforests in Central America. The female lays a bunch of eggs into a small waterhole. During hatching, the male aggressively defends the territory. He also takes care that the waterhole does not dry out by bringing water inside his cloaca to supplement evaporated water. Small tadpoles hatch after 10 days. Mother picks up every one of them individually and transports them to the individual pools of water. This separation of infants is necessary as tadpoles are very aggressive. They would attack each other, so they must be physically segregated if the mother wants to ensure the survival of more than one tadpole. Even after she completes the transport and finds a suitable watery hole for each of her babies, her work is not done. Tadpoles need food, and the mother also provides this. She visits every child every day or every few days and lay some unfertilized eggs into their pools to provide them with food. All this time, her mate keeps the watch and guards the babies against the predators.

Caecilians look like giant worms, but they are amphibians, living in rainforests around the globe. They spent most of their time underground. Yet, mothers take great care of their young, virtually sacrificing parts of their bodies for the well-being of their younglings. Baby caecilians feed on her mother's body. In some species, young caecilians hatch inside the mother's oviduct, and they consume cells lining the oviduct. In other species, they hatch outside the mother's body, but young caecilians feast on their mother's skin. Mothers develop a thickened epidermis, rich in nutrients around their backside, and babies feed on this skin every third day as the skin needs 3 days to regrow.

Numerous animals are taking great care of their young ones, either one or both parents. However, strong contenders for being the champions of parental devotion would undoubtedly be Adelie penguins. Male and female Adelie penguins often mate for life. Although they separate each year during their long travels along Antarctica in search of food, they meet again year after year and reproduce in the same place. Adelie penguins build their nests from stone only on firm land, but they can only find food in the unfrozen sea. In the summer, this is reasonably close to their breeding grounds. But during the harsh and long Antarctica winters, ice covers vast stretches of the ocean. The open sea is far from the firm ground. Therefore, penguins have to migrate for many miles between winter and summer months to be close to the open sea and obtain food.

Each autumn, when days start to shorten, Adelie penguins migrate for several hundred or even thousand miles in search of food that is only available in the unfrozen sea. As the winter wanes and days lengthen, this is a signal for penguins to make the same trip back. They walk to their breeding grounds on ice-free shores where males and females, mates from the previous season, usually meet at their nests if they are both still alive.

Males generally arrive a short time before the females, and they begin to work on their nests immediately. Those with old nests from the previous season rush to repair them as winter snow and winds might have dislodged the stones from which the nests are made. Young males build new nests in anticipation of females. But rocks are a precious commodity in Antarctica, and in penguins' breeding grounds, there is often a scarcity of stones. This forces male penguins to steal the rocks from each other to build a nest that will impress the female and later protects their eggs. But the looting of stones from each other is not taken lightly and often leads to serious fights.

Once ladies arrive, things are mostly settled down between the males, and the mating rituals could begin. Old pairs find each other, and new males try to impress single females with their nests and lure them into mating. After eggs are laid, males and females take turns in incubating them. They make sure that eggs are not left unattended even for short periods. As little as 5 min of exposure to the cold Antarctic climate is enough to destroy the eggs. Yet, penguins have to feed during the hatching, which usually lasts 36 days. Therefore, excellent cooperation between males and females is needed, coordinating who will take care of the egg and who will forage for food. The partnership between males and females is so crucial that females know that they cannot hatch the eggs and provide for a chick alone. If a male mate dies for any reason, a female will usually abandon the eggs, knowing it would be futile to care for the eggs alone.

After hatching, mum and dad continue to watch over their baby for about 4 weeks, when all small penguins are put together in some kind of crèche. Only then can both male and female finally go to the sea along with each other. They can now stop fasting periods that lasted for about 2 months, which is not a small feat in Antarctica's cold, harsh environment.

We often say that the mother's love is the only unconditional love that exists. As we have seen, a strong bond between parents and their offspring is not something unique to us. It is present in many animals, not only in our closest relatives, mammals. Mothers and sometimes fathers sacrifice themselves to bring up their family, protect their younglings, and ensure the propagation of their genes into the next generation. It is difficult to say whether caring for offspring by frogs, birds, or rats and elephants is something we can call love, or it is just a basic instinct helping to preserve one's genes. But, as we will see in later chapters, some mechanisms controlling parental behavior in males and females are similar to what is happening in our bodies and brains when we love. So, it is very likely that feelings of love and attachment originate from the motherly or fatherly love for their offspring and that early origins of love exist in animals showing maternal affection.

# Chapter 15
# Depraved Hooligan Penguins

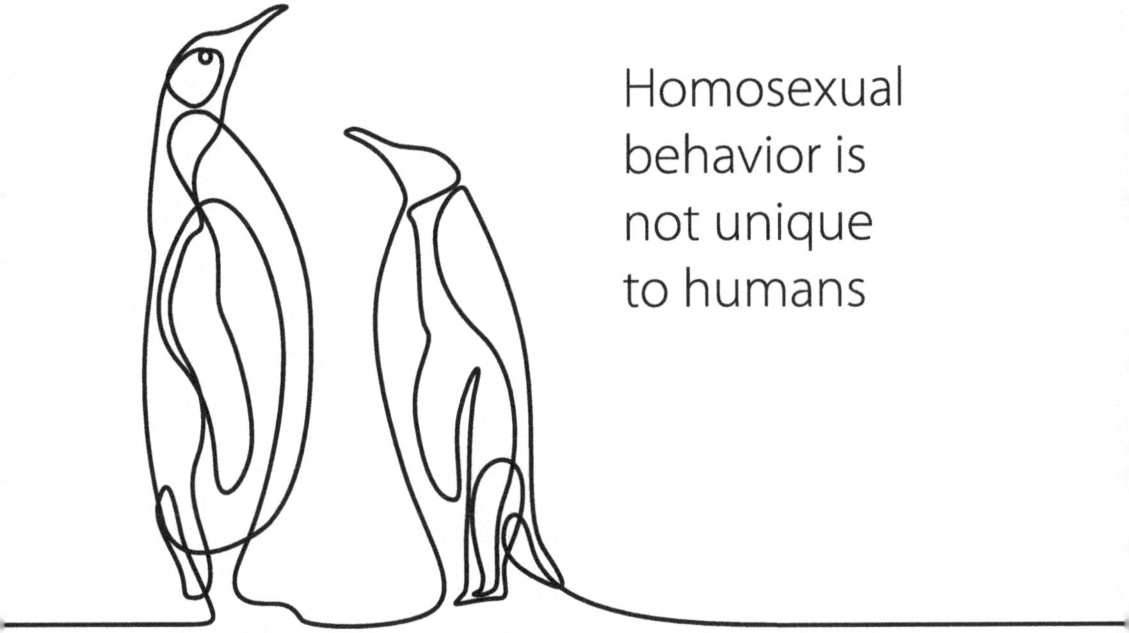


*Different couples seem to vary much in character or mood. Some can be seen quarrelling violently, whilst others appear most affectionate, and the tender politeness of some of these latter toward one another is very pretty to see.*

George Murray Levick: Antarctic Penguins

In 1910, avid biologist George Murray Levick joined the expedition of Captain Scott to the south pole as a medical officer. Besides his official role, he used his time with the expedition for biological observations. In particular, George Murray Levick was interested in penguins. He was the first person to observe and study penguins throughout the entire season when he stayed with a penguin colony in Cape Adare in Antarctica. Levick was a keen observer of these cute animals, yet, he was shocked by some behaviors displayed by the penguins. In his notebook, he wrote observations of what he called "depraved sexual acts of hooligan males mating with dead



G. Majdic, *Soul Mate Biology*, Fascinating Life Sciences,
https://doi.org/10.1007/978-3-030-67212-6_15

females" in Greek, so nobody but other scientists could read them. George Murray Levick observed Adelie penguins engaging in many sexual behaviors that were just too much for the sensible soul of a scientist from the beginning of the twentieth century. He noted that males occasionally mate with dead females, sexually abuse young chicks, and even have same-sex sexual encounters.

When Levick returned to England, he thought about publishing all his observations on penguins in the scientific paper but was persuaded by his peers that this would be just too much for society. So, he published an important article about penguins' behavior without the chapters about their sexual "depraved" acts. Fortunately, Sidney Harmer, a keeper of Zoology in the Natural History Museum, kept several copies of the section on the penguins' sexual behaviors. Two copies vividly describing homosexual behavior in penguins were thus preserved until the present days.

These notes are, by far, not the oldest notes about homosexual behavior in animals. Aristotle already described more than 300 B.C. same-sex mating in pigeons and partridges and focused particularly on spotted hyenas. Spotted hyenas are curious animals. Females are masculinized as they are exposed to high levels of male sex hormone testosterone during development. It is not known why and exactly how this happens, but both during development in the womb and later in life females have high testosterone levels in their blood.

Consequently, female spotted hyenas are very aggressive. Females are much more aggressive than males, which is unusual in animals. As testosterone is a hormone that causes the development of the penis from indifferent primordium, female spotted hyenas also have clitorises that look more like penises. They mate through these pseudo-penises and give birth through them, so it seems the evolution quite complicated hyenas' love lives. It is quite a formidable task for a male hyena to find the entrance into the clitoris and push his penis inside a very similar organ. Even inside, the female sexual organs are very long and convoluted and pose a challenging mission for a sperm. Perhaps this long and twisted path is a reason for the evolution of such unusual sexual organs. Maybe this is a sexual selection at its finest.

One hypothesis why spotted hyenas have such a complicated reproductive system suggests that this long and arduous path for sperm ensures that only the best and the fittest sperm reaches the egg at the end of the labyrinth. Furthermore, a female can change her mind about her beloved after the mating. By urinating, she can flush the semen out of her tract if she had second thoughts about her mate after copulation. Another simpler hypothesis suggests that complicated reproductive tract and penis-like clitoris have just evolved as an undesirable side effect of evolutionary increasing aggressive behavior, mediated by increased testosterone production in females. In addition to the penis-like clitoris, female spotted hyenas are usually larger than males. As male and female spotted hyenas' external appearance, especially their sex organs, are so similar, it seems they are engaging in the same-sex intimate act when they are mating, as described by Aristotle.

Although homosexual behavior does occur in hyenas, it is not as common as in some other animals. Earlier observations and descriptions of same-sex mating in spotted hyenas were mostly sexual encounters between male and female hyenas.

One might expect that there would be more same-sex sexual encounters in animals where females are so heavily masculinized. However, it seems there is some very fine-tuning going on regarding the masculine development of the hyena's female brain. Areas of the brain that usually differ between males and females in many animals, and are known to vary because of testosterone, are different between sexes also in spotted hyenas. This presence of sex differences in the brain suggests that the female hyena brain is somehow protected from high testosterone levels and at least partially develops as a female brain. This is entirely unlike most other mammals, where testosterone masculinizes the female brain if present in the body in the period around birth. Although hyenas' bodies are heavily masculinized, their brains are not. Consequently, their mating instincts are as expected for each sex.

When we think about homosexuality, most people think of humans. Many believe it is something that evolved during civilization and is not present in wild animals. The truth is very different, and homosexuality is widespread in many animals, from insects to mammals. This does sound out of the box and surprising, as it does not fit well with the general theory of evolution. According to evolutionary theories, the primary purpose of reproduction is to pass the genes onto the next generation as efficiently as possible. Love and being in love are most likely part of the evolutionary trick to force us to reproduce. Genes that would counteract the ability to reproduce, possibly by making a male or female homosexual, should disappear from the gene pool. Homosexual individuals do not reproduce, at least not as efficiently as strictly heterosexual individuals, so genes promoting homosexuality should be slowly eliminated from the population. Nevertheless, homosexual encounters are relatively common in the animal kingdom and can be observed in many animals.

Many years of observation of Japanese macaques, cute monkeys, warming themselves in hot springs during snowy cold winters in Japan, revealed that females often engage in homosexual relations with other females. Females mount other females, rub their genitals together, and, judging from their facial expressions, enjoy these acts. Usually, such females stay intimately together after the sexual encounter, cuddle together, and stroke each other on their fur, offering mutual protection. However, as often in animal homosexual behavior, these Japanese macaque females are not strictly homosexual. The same females that mount other females (or let other females mount them) mate with males to become pregnant and have babies a few hours or days later. Yet, female homosexual behavior is so common that it cannot be a mistake or a chance.

We do not know the reason behind this homosexual behavior, but some scientists have their suspicions. In Japanese macaques, females seem to be much better lovers than males. They are more versatile in choosing sex positions. They use a much larger variety of movements during sexual encounters. With these superior abilities, females often initiate sex, mount males, and try to arouse them. As they are very active trying to stimulate the males and enjoy the sex, perhaps the homosexual behavior just spilled over during evolution. Females became indiscriminate about who to mount to derive the pleasure.

But it could also have an evolutionary advantage. If males are slow and less interested in mating than females, females need to arouse males as efficiently as possible. The "training" with other females could perhaps improve their sexual behavior and skills. Females with a lot of training, regardless of whether with males or females, could have an advantage. They are better, more experienced lovers and surpass inexperienced females in convincing males into the mating, thus increasing their chance of transmitting genes into the next generation.

Homosexual behavior can be observed even in such animals like the fruit fly. Fruit flies are regular participants in biological research due to the ease of manipulating them. Their short life expectancy means that many generations can be observed in the laboratory. Thus, evolutionary processes and rules of heredity can be studied. They are probably one of the most studied organisms on earth, and scientists are well acquainted with their genes and behavior.

When fruit fly goes past the metamorphosis and develops into an adult male, males seem to be incredibly horny. They would try to mate with any other fruit fly around, regardless of sex. Again, they are not strictly homosexual but rather bisexual and only for a limited time. After a particular time, males seem to learn who is who regarding the sex and who would be a better mate to produce offspring and pass genes onto the next generation.

So why are initially male fruit flies so horny and indiscriminate? As with many animals, pheromones have a crucial role in the reproduction of fruit flies. Male fruit flies recognize the females by sniffing their pheromones. However, the pheromones secreted by the females could differ somewhat regarding fruit flies living location and what food is available to them, just like our pee smells differently depending on the food we eat. Males might initially have problems recognizing the females. Therefore, indiscriminate bisexual behavior in male fruit flies might have evolved, so males would not miss a potential female mate.

The rule for the young fruit fly males seems to be mount as many mates as possible until you learn well who is female and who is not. This would undoubtedly have an evolutionary advantage. A male fruit fly, who would mount every fruit fly passing by regardless of sex, will not miss an unimpregnated female, even though he will have few misses by mounting males. He will, therefore, increase his spread of genes. On the other hand, a choosy male that would look strictly for specific pheromonal smell might miss some females and therefore the opportunity to pass his genes onto the next generation.

Besides penguins, many birds engage in homosexual behavior, not just during sexual encounters but also in living and raising children together. Such behavior is well known in Laysan albatrosses living on Hawaiian islands. In these birds, the sex ratio is often skewed, usually with females exceeding males by an unusually large margin. In such populations, female-female couples often form, and such female-only couples behave just like male-female couples.

In albatrosses, parents take care of the young for a long time, usually longer than 225 days. One pair could care only for one chick, although single females could lay and hatch eggs. Therefore, for a species with surplus females, it makes sense that birds establish same-sex couples capable of rearing and caring for orphan chicks, as more chicks in the colony could be brought to life. Indeed, these female-only couples are often good parents.

Although male albatrosses spend time caring for chicks with only one female, they are willing to have occasional sexual encounters on the side when their madam is not looking. Single females seduce engaged males into mating, but they do not look back after sex. Such single fertilized females find the female partner. They lay eggs and care for the chicks with their female partners, and they are as good at bringing up the children as any heterosexual couple. From the evolutionary point of view, this is undoubtedly a win-win situation. The best males could mate with more than one female, and they could transfer their genes into the next generation through more than one chick per year.

Similar homosexual behavior is present in black swans. Up to a quarter of pairs might be same-sex marriage, but with two males. Male swans could not lay eggs even if they would mate on a side as female albatrosses do, yet, these male pairs do have parental instinct and want to be fathers. To get their children, they sometimes engage in dirty tricks. Quite often, one partner from a single-sex marriage finds a female. He lures her into the mating, wait until she lays eggs, and then, very unchivalrous, chase her out of the nest. He then takes over the eggs and later cygnets with his true love male partner.

Other male couples find a female that already has a nest with laid eggs, chase her away from her eggs, and acquire the nest with the eggs. Afterward, they behave like excellent parents, make sure that eggs hatch, and take good care of their fostered cygnets. It seems that male pairs are even better at taking care of the chicks than heterosexual couples. Perhaps not because they are better parents but because males are stronger. Two males could more easily defend the chicks from the predators. Yet, even with albatrosses and swans, despite single-sex parents being excellent parents, the individual's behavior is not strictly homosexual or heterosexual. It is a response to the skewed sex ratio in a confined population. Female albatrosses engaged in same-sex parenting will be happy to engage in male-female parenting a year later.

In mammals, homosexual behavior can be seen in many species, both domestic and wild animals. Male-male sexual encounters are relatively common in giraffes, lions, elephants, dolphins, bison, and domestic cattle. In cattle, bulls readily mount other bulls, and cows during estrous, stimulated by increased estrogen levels in their bloodstream, often mount other cows. So many mammals do engage in homosexual encounters. However, what is the biological drive and the purpose of such homosexual behavior is not clear.

Our closest relatives, bonobos, are very sexual animals. They engage in sex so often that they are sometimes called "our oversexed relatives." They use sex in all sorts of social interactions. They are the animals that most clearly do not engage in sex for reproduction only. Homosexual encounters are especially frequent between females, where up to 60% of all sexual encounters could be homosexual. Yet, again, there are no strictly homosexual individuals.

All individuals, males and females, engage in homosexual and heterosexual encounters. Bonobos often use sex to ease the tensions between the two individuals. For example, a jealous male bonobo might chase away a male, approaching his female. Yet, instead of fighting, such male competitors often sit together after the chase, rubbing each other's genitals. A female might hit a juvenile from another female. Yet, the infant's mother would often approach an aggressive female and engage her in genital rubbing instead of entering the fight. Bonobos also use sex to advance their position in the group. Younger individuals are engaging in sex with more dominant males or females, thus promoting their social status.

All these examples of homosexual behavior in animals are always bisexual behaviors. In none of the species described, animals are strictly homosexual. Besides humans, the only animals that seem to know strictly homosexual individuals are domestic sheep. Up to 8% of rams display strictly homosexual behavior. They always mount only other rams, even when given a choice of ewe or ram.

Interestingly, the hypothalamus, which is usually sexually dimorphic, differs between heterosexual rams and homosexual rams. Certain groups of cells in the hypothalamus that are generally larger in males than in females are in homosexual rams more like in ewes than in heterosexual rams. It also seems that homosexual rams have lower levels of testosterone and higher estradiol levels in their blood, what could be connected to their behavior.

Why would strictly homosexual rams exist if this apparently could not have any evolutionary advantage for the species? One plausible hypothesis suggests that the same genes that make rams homosexual increase fertility in ewes. If this is true, this could have sense, as more fertile females will produce more offspring and pass their genes more often into the next generations. As only one male is needed to impregnate a large number of females, genes that would increase the fertility in ewes would be propagated to the next generations, even if the side effect of such fertility genes is homosexuality in some males. Due to a large genetic pool in any population, there would still be enough heterosexual males at any given moment to ensure the proper reproduction and continuation of the species.

Sexual encounters between same-sex individuals are relatively common in the animal kingdom. But in almost all animals, this is not strictly homosexual behavior, and individuals would engage in homosexual and heterosexual encounters. So, it is likely that such homosexual behavior is just a spillover of the evolutionary processes encouraging sexual engagements to increase the chances of survival of the species.

Reproduction is the most fundamental process in all living organisms as it ensures the propagation of the species. Therefore, active and frequent reproductive activities are encouraged by evolutionary processes. Homosexual behavior is, therefore, likely to be just a side effect of selection for a large sexual drive that would ensure the continuation of the species. Although some animals like bonobos use homosexual relations to advance their social status and strengthen the bonds within the group, these are not love and attachment. They are rather an overly active utilization of the evolutionary advantageous powerful sex drive. Therefore, homosexual behavior in animals does not seem to be connected with love as it is in humans.

We still do not understand the biological and evolutionary explanation for homosexuality, either in animals or humans. Although levels of the male sex hormone testosterone seem to be lower in homosexual than in heterosexual rams, testosterone levels in homosexual men seem to be higher than in heterosexual men. In homosexual rams, some brain characteristics are more female-like than male-like, but this is not the same in humans.

So far, no structural differences in the human brain that could overlap with homosexual behavior have been consistently found. Some studies suggest that homosexuality is more common in men and women exposed to inappropriate levels of sex hormones before birth, but this is difficult to study and confirm. We do not have definitive answers yet whether there are any biological roots for homosexuality. It is likely that, as in all other characteristics, genes, hormones, and experience interact to induce same-sex preference, and there is no single biological path to homosexuality.

Homosexuality is present in many animals and is not something unusual in the animal world. We do not understand what regulates sex preference, but heterosexual and homosexual relationships are widespread in nature. Therefore, we should accept that love exists in many different forms, and homosexuality is not something unique to us, humans, but is also present in the animal world.

# Chapter 16
# Belligerent Elephant Seal Landlords

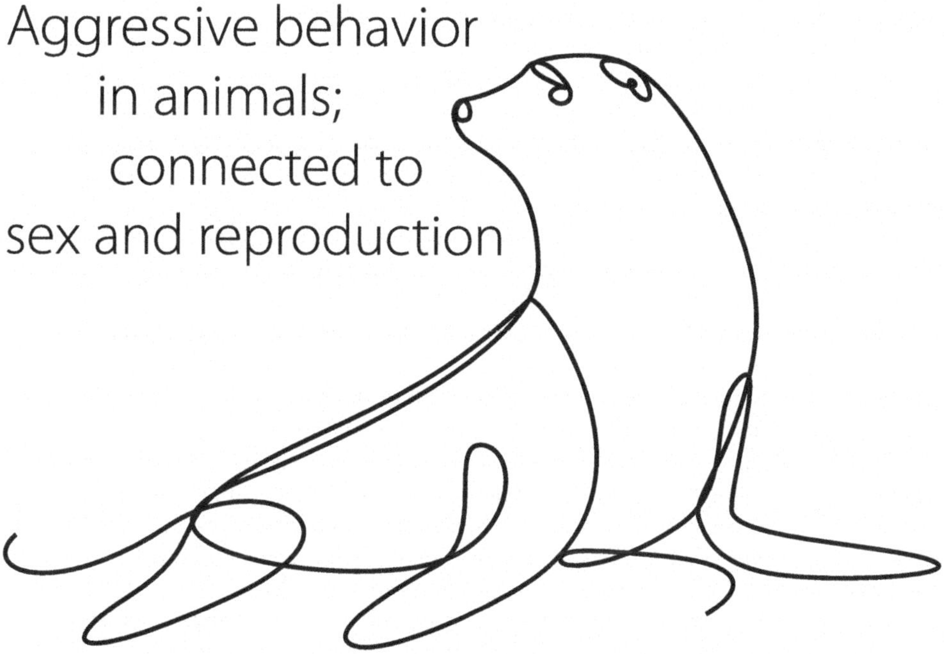


*The desire for magic cannot be eradicated. Even the most supposedly rational people attempt to practice magic in love and war. We simultaneously possess the most primitive of brain stems and the most sophisticated of cortices. The imperatives of each coexist uneasily.*

Erica Jong: What Do Women Want?

Scar, Simba's uncle in the Lion King cartoon, was undoubtedly aggressive. But so was Simba when it came to defending his family and his friends. But unlike Scar, Simba is perceived by most of us as friendly and cuddly. Scar, Simba, and Simba's father Mufasa were portrayed very differently in the popular Disney cartoon. While Simba and Mufasa are kind and caring, Scar, Mufasa's brother, is insidious, cunning, and violent. Yet, they all behave aggressively when the need arises. They fight fiercely and violently. Aggression is an incessant part of nature and is present in



G. Majdic, *Soul Mate Biology*, Fascinating Life Sciences,
https://doi.org/10.1007/978-3-030-67212-6_16

many different forms. As seen in The Lion King, even likable characters are sometimes aggressive when they have to obtain or defend something, usually the territory to live in or the mate to create the progeny.

"War is like love, it always finds a way," said Bertolt Brecht, and if we substitute the word war with the word fight or brawl, this is true for humans and animals. Although animals, with the probable exceptions of chimpanzees, do not wage wars, fighting and aggression are ever present in the animal world.

Many animal males, including our own, are frequently aggressive when they want to obtain real estate or win the loved one's heart. Often, these two missions are not separated. The prime location offering food and protection attracts more females. Beachfront property for an elephant seal or a large villa on Beverly Hills for a human male has similar effects on the opposite sex. They increase the chances of winning the heart of the loved one.

Combats for power, possessions, and brawls to win the heart of the loved one are integral parts of the evolutionary path, a constant struggle for successfully transferring one's genes into the next generations. Likewise, moms being aggressive when defending their brood has essential implications for protecting children and the survival of the genes into the next generation. These aggressive behaviors are intimately linked to the broader concept of survival of the species. It is, therefore, not surprising that similar signaling molecules regulate both love and aggression.

Everybody witnessed brawls between teenage boys over the girls. Numerous duels have been fought over women in history. In one of the most famous stories of all times, a war started because of love, when Helen of Sparta was abducted by Paris from Troy, leading to the famous Trojan War. In all such events, the same hormones that make us long for love also elicit aggressiveness and violence.

There are many different kinds of aggression. Often, the word aggression is used very broadly. But in a strict biological or psychological sense, aggression is defined as behavior aimed at hurting somebody who does not want to be hurt. Aggression is also often used interchangeably with violence, although aggression and violence differ. Aggression can be verbal, assuming a threatening position, or sending a bullying message without physically hurting somebody. Violence, on the other hand, is defined as an aggressive act aimed at hurting somebody physically. So, violence is aggression, but not all aggressive behavior is violent.

Climbing up the corporate ladder can be very aggressive, but it is usually not violent. Children calling each other names are aggressive, but they are not violent until they begin to fight and hurt a classmate physically. Dog baring its teeth and barking is undoubtedly aggressive, but often such dogs are not violent and do not attack humans or other dogs, as it goes in the saying—barking dog doesn't bite. In the animal world, violent behavior has a different meaning. Carnivorous predators are unquestionably violent when they hunt and kill their prey, but they have to do this to provide food.

Such behavior is part of the natural cycle, essential for immediate survival. Lioness violently attacking zebras in the African savannah certainly appears violent, but she does not act with the intention to hurt zebras. It does it only to provide the provisions for herself and her kin. Buffalos defending themselves from a lioness act violently, but again, not with the conscious aim of inflicting physical pain on the lioness. They just protect themselves. Both these behaviors are very violent for the observer, but they are part of the general survival instincts of the animals.

Hostilities between the members of the same species, however, are different. They are not momentary basic instincts of the individual survival. Aggressive behavior between the members of the same species is associated with one of two objectives—defending or acquiring the living space (what includes constant access to the food in one's territory) and obtaining females to mate with. Therefore, aggressive behavior between members of the same species is connected with the survival of the species, just as mating and love are. Miguel de Cervantes wrote in the famous novel Don Quixote: "Love and war are the same things, and stratagems and policy are as allowable in the one as in the other." De Cervantes did not know anything about the hormones. Yet, if he would, he could have concluded that some of the same signaling molecules also regulate love and war in our and animals' bodies.

With elephant seals, it is all about the real estate. Elephant seals nest on the shore, but they feel much safer in the water. This is especially true for the younglings, which are often targets of the predators living on the land. Therefore, the beachfront property is the most valuable among elephant seals. The mother and her pup can quickly escape into the ocean and avoid danger if they nest close to the sea.

Elephant seal males are very territorial animals, and they fight fiercely for prime locations. The strongest and the fiercest fighters acquire the best real estate, a lot on the beach closest to the sea. Such lots are not just the most romantic spots, offering the best views of the sunset over the sea or eavesdropping on the gentle murmur of the ocean. The land closest to the sea brings an extra prize to the sires—a chance to mate. Elephant seal females are not landowners. They are looking for potential partners, and just as we like beachfront cottages, so do female elephant seals like lots closest to the sea. Females choose for partners the strongest males, who are also the proprietors of the best beachfront residences.

As elephant seals are not monogamous, the best real estate on the beach attracts multiple females. The strongest and fiercest warriors, therefore, receive the prize in the form of numerous lustful females. Weak males with lots further away from the sea don't get a chance to mate, and therefore, they cannot transfer their genes into the next generation.

Male elephant seals are much larger than females, and they are very aggressive during the mating periods. There are numerous fights between males, and many end with the death of the weaker opponent. However, being a winner comes with a price. Although the dominant male could sire many, sometimes up to 150 younglings in one season, such dominant males usually die 1 or 2 years after their peak reproductive year. Numerous fights to defend their territory and caring for multiple females deplete the energy of dominant males and shorten their life span. This short life span of dominant males gives other males the chance to become dominant and ensure that various combinations of genes are transferred to the next generations. So being the strongest and the most aggressive male brings the prize, but it comes with a deadly cost.

Aggressive behaviors between the members of the same species are not directly connected with the hunger and immediate survival but with more complicated evolutionary survival strategies of the species. Occupying and defending the land from intruders provides access to food sources for a long time, not just for a single dinner. Wolves pack aggressively protecting their territory from the interlopers ensures plenty of food for all the pack members, including cubs and elderly wolves. Similar territoriality to defend food sources is present in many animals, not just mammals but also birds, fishes, and even insects.

Owning a land also provides better access to females, as in elephant seals. The ultimate prize of the evolutionary game, mating with a female and transferring the genes into the next generation, is reserved for the winners either for the winners in the struggle for land or the champions of combats. In both cases, the winners are conquering the hearts of the females. In nature, most often, only the strongest, the loudest, or the smartest members of the species get a chance to mate and sire a progeny. Evolution in many animals favored the most aggressive males to be the most successful fathers. Hence an explanation for the ever-present antagonistic behavior between males of the same species and the intimate connection between love and aggression.

Occupying and defending the land and fighting for a mate are most often left to the males. However, in some animals such as African lions or spotted hyenas, the madams hold the reins within the pride. Females of most other mammals are usually more docile, although they can be very aggressive in a different context.

Aggressive behavior often increases in females when they become mothers and have to defend their brood. After females become moms, they turn from placid girls to fierce warriors. They are prepared to risk their lives in the protection of their brood. They protect the infants from predators and from the members of the same species that might want to harm their babies. Many dog breeders know how bitches change their character after becoming moms. From calm and gentle pooch, they become highly protective and sometimes aggressive beasts. They might act aggressively toward the owner despite the previous lovable relationship. They often chase away the father of the puppies as they may perceive him as a threat.

Ultimately, fighting for the territory, fighting for access to females, and defending the children are all part of the broader strategy to transfer one's genes into the next generation, what are also the aims of reproduction, sex, and love. Aristotle observed more than 2000 years ago that aggression and mating are intimately connected: "The statements made in regard to the pairing of the sexes apply partly to the particular kinds of animal and partly to all in general. It is common to all animals to be most excited by the desire of one sex for the other and by the pleasure derived from copulation. The female is most cross-tempered just after parturition, the male during the time of pairing; for instance, stallions at this period bite one another, throw their riders, and chase them. Wild boars, though usually enfeebled at this time as the result of copulation, are now unusually fierce, and fight with one another in an extraordinary way, clothing themselves with defensive armor, or in other words deliberately thickening their hide by rubbing against trees or by coating themselves repeatedly all over with mud and then drying themselves in the sun.

They drive one another away from the swine pastures, and fight with such fury that very often both combatants succumb. The case is similar with bulls, rams, and he-goats; for, though at ordinary times they herd together, at breeding time they hold aloof from and quarrel with one another. The male camel also is cross-tempered at pairing time if either a man or a camel comes near him; as for a horse, a camel is ready to fight him at any time. It is the same with wild animals. The bear, the wolf, and the lion are all at this time ferocious towards such as come in their way, but the males of these animals are less given to fight with one another from the fact that they are at no time gregarious. The she-bear is fierce after cubbing, and the bitch after pupping."

Many male animals exhibit enhanced aggression around the time of breeding and females after they deliver their brood. Therefore, it is not surprising that similar hormonal and brain circuits that control sexual reproduction also regulate aggressive behaviors. These mechanisms ultimately evolved into love in our species. Although love and aggression seem to be completely antagonistic behaviors, they are biologically related. Notwithstanding, unfulfilled love often turns into aggression and even violence. Considering all these, it is quite natural that one of the primary mediators of aggressive behavior in males is the same substance responsible for the lust in men, male sex hormone testosterone.

During most of the year, stags live peacefully together in bachelor groups, grazing in the vicinity of hinds. However, when day length starts to wane and autumn approaches, their antlers grow and harden. At the same time, peaceful stags turn into highly aggressive brutes. They meet each other in the clearings in the woods and fight fierce battles. Many stags are injured, and some even die during these bouts. The winners receive their prize, mating with a willing female. But once all females are impregnated, when days grow even shorter, temperatures fall and winter approaches, stags shed their antlers and return into a peaceful co-existence within the bachelor herds.

Changes in the behavior and testosterone levels in stags' blood are closely correlated. At the end of the summer, when stags' internal clock inside their brains detects shortening of the day, their testes become active, and they start to produce more testosterone. Testosterone causes growing antlers to harden, turning them into a formidable weapon. Testosterone also changes the docile bachelors into bellicose warriors. Once stags have completed their reproductive obligations, the brain sends another signal to the testes and testes lower testosterone production. Antlers are shed, and stags become serene again.

There are other indications that testosterone has an essential role in the regulation of aggressive behavior. Although boys or young male animals often engage in rough plays, they do not become aggressive until puberty. Prepubertal stags with low levels of testosterone do not participate in fights between mature males. In most animals, with few exceptions, females are less aggressive than males. Although mothers can be very aggressive when protecting their offspring, females are, in general, more peaceful than males. Furthermore, castration in many animals reduces aggression.

This has been known since prehistoric times and was and still is used with domestic animals. The neutered horse is much easier to handle than a stallion, full of testosterone. The adult bull is hard to control, but the placid ox can be easily harnessed in the yoke and is willing to pull the plow. And finally, if we give the castrated animals back testosterone, they usually regain aggression.

Most animals mate and have offspring in warmer months when the chances for survival are the highest. Many animals are less aggressive during the wintertime. In the harsh season, the resources are scarce, survival is threatened, and fighting could endanger the survival of the individuals as they would spend too much energy.

Meadow voles occupy the grasslands during the summer when food is plentiful. The summer is also their breeding season, and testosterone flows high through the vole males' veins. During warm weather, meadow voles living in grassland and red-backed male voles living in nearby woods are intolerant to each other. They occupy mutually exclusive territories and are highly aggressive toward each other if they meet by chance. However, during the winter, meadow voles move to the woods, seeking protection of the trees. They intrude into the territory of red-backed voles. Interestingly, members of both species, highly aggressive to each other during summertime, cohabitate peacefully in the forest throughout the long winter months. During the cold, dark winter, the voles also don't mate. Their testes produce less testosterone, and this allows the peaceful cohabitation of the two species.

Clearly, testosterone is the main culprit for aggressive behavior. So, can we reduce the testosterone in men and we will all be living peacefully? Unfortunately, it is not that easy. Although there is a correlation between aggressive behavior and testosterone, this correlation is not entirely straightforward.

Reduction in aggressiveness during the non-breeding season or after castration is not universal. In some animals, such as Mongolian gerbil, neutering does not reduce aggression. Both Syrian and Siberian hamsters are more aggressive during the winter, despite low testosterone levels produced by their testes. In hamsters, it seems that another male sexual hormone, dehydroepiandrosterone, produced by adrenal glands, is causing increased aggression during the winter when reproductive organs are shut down. When days are short, blood levels of melatonin, a hormone produced in the dark and intimately connected with measuring seasons, increase. Melatonin stimulates the adrenal gland to produce more dehydroepiandrosterone, causing increased aggressiveness in hamsters during the winter. Therefore, it is not all about testosterone. In its absence, other similar hormones might have comparable effects. In general, aggressiveness in numerous animals varies with seasons and hormones, as do reproductive instincts.

We are not seasonal animals, as we can have sex and reproduce at any time of the year. However, there are annual changes in the levels of hormones also in our bodies. In the summer, men have more testosterone in their veins.

Although the link between higher testosterone levels in summer and violent behavior was not directly explored, there is evidence that men are more aggressive in the summer. There are more violent crimes happening in the summer, and problems with the inmates in jails are more common in the warmer months than in the winter. Of course, correlation is not causation, and this increase in violence might simply arise due to more socializing during summers. However, there is some evidence that domestic violence also increases during the summer months. This seasonal difference cannot be explained by more socializing as families usually spend more time together at home during winters than on sunny summer days. We certainly cannot conclude that higher testosterone levels during summer months cause an increase in violence in men, but the available evidence indeed points that way.

Testosterone acts on the brain in different ways and at different times. During development, testosterone triggers some permanent changes in the brain, consequently causing behavioral differences between males and females. In rats and some other rodents, increased aggressive behavior in males is preprogrammed by the early exposure to testosterone. If male rats are castrated immediately after birth, they will not be aggressive as adults, even if given large amounts of testosterone in adult life. On the other hand, male rats castrated in adulthood will be aggressive if they receive testosterone. And finally, if we give female rat testosterone immediately after birth and again in adulthood, such female rats will be much more aggressive than ordinary females or females that receive testosterone only at birth or only in adult life.

All this clearly shows that in rats, the testosterone's presence early in life and in adulthood is necessary for aggressiveness. However, in some other animals, puberty is the main period for turning on brain aggression centers. Mongolian gerbils and hamsters are not aggressive if they are castrated at any time before puberty, not just within a few days after birth. So clearly, while brain aggression circuits are organized in rats immediately after birth, in Mongolian gerbils and hamsters, this happens only during puberty. In many animals, adult testosterone is necessary to control aggressive behavior, but this is not universally true. Some animals, like dogs, can remain aggressive after castration. This is also true for some primates. Male rhesus monkeys remain aggressive even after neutering. Female rhesus monkeys are aggressive if their mothers were treated with testosterone while pregnant without the need for testosterone in adult life to display aggressiveness.

So, are we, humans, aggressive because of testosterone? This question is undoubtedly challenging to study, but it seems that the partial answer to this question is yes, but not universal yes. Girls exposed to higher than normal testosterone levels prenatally due to rare genetic defects more often engage in rough games than their sisters who lived in the same households and were not exposed to the elevated levels of testosterone before birth.

There are correlations between blood testosterone levels in men and violent behavior, but these correlations are relatively weak. On the other hand, doping with androgens in male athletes does seem to increase aggressiveness. It seems there is a role of testosterone in human aggression, but testosterone is definitely not the only culprit. Testosterone is probably more an enabler than a trigger of aggressive behavior. This could explain substantial differences in aggressiveness between people. We all know that some people respond aggressively after minimal provocation, while other men never engage in fights. Yet both groups have similar levels of testosterone in their blood.

One crucial factor in the display of aggressive behavior is experiences, not only in humans but also in animals. Although testosterone is obligatory for the proper expression of aggressive acts in mice, mice behave differently after castration regarding their previous experience. If males never engaged in aggressive bouts between males, they will not display any violence toward other males after castration. However, if males were experienced fighters, they retain some aggressiveness even after castration, although at lower levels than males with the normal levels of testosterone.

Testosterone is, therefore, essential for displaying aggressive behavior, but it is neither sufficient nor exclusive regulator of aggressiveness. It is probably a facilitator, but not the inducer of aggression. We cannot blame the aggression on testosterone. Although testosterone likely enables our aggressive brain circuits to act, other brain centers in our conscious brain have the power to overcome the testosterone switch. Therefore, violent behavior cannot and should not be blamed on testosterone.

In addition to combats between males, there is another type of aggression connected with love. Mothers of all mammals take great care of their offspring. They nurse them, warm them, and cuddle with them. But if a threat arises, most mothers will protect their children, whatever the costs. We can read the stories about brave mothers saving their children from fires, storms, or wild animals. Although it is challenging to study maternal aggression aimed at protecting children in our species, there is ample evidence from the animal world that mothers are more aggressive when they defend their brood.

Female mice and rats are generally docile animals, but they are highly protective when they are mothers. They will fight any male recklessly wandering into the vicinity of their nest. So are the bitches, monkeys, apes, and most other mammalian moms. This increased aggression in moms must be linked to hormones, as it only emerges during the pregnancy.

One of the essential hormones for maternal aggression in mice is progesterone, a hormone similar to estrogens. Ovaries secrete modest amounts of progesterone during the second half of every menstrual cycle. However, if a female becomes pregnant, levels of progesterone rise sharply. Initially, it is produced by the ovaries, while later during pregnancy, the placenta is the primary source of this hormone.

Progesterone has essential roles throughout the pregnancy. It prepares the womb for accepting embryo by stimulating uterine glands to produce early food for the embryo. It reduces cramps of the uterus that would otherwise cause the expulsion of the baby. At least in mice and rats, progesterone has another role during late pregnancy as it triggers aggressive behavior. If the action of progesterone is blocked during the late stage of pregnancy, such animal moms are less aggressive.

But just like testosterone in males, progesterone alone is not the only factor regulating increased maternal aggression. As progesterone's primary source during pregnancy is the placenta, progesterone levels drop immediately after birth, when the womb discharge the placenta. Yet maternal aggression remains high for several weeks in mice and rats and months or even years in other animals. The nursing of babies and the presence of babies seem crucial for maternal aggression in animals.

If mouse or rat mothers are separated from their babies when they meet an unacquainted male, they are not as aggressive as if they encounter an intruder in the vicinity of their nests.

What other hormones regulate aggressive stance when moms have to defend their children is not yet known. There is likely some role of hormones involved in other aspects of love and attachment, such as oxytocin, vasopressin, serotonin, and possibly stress hormones. Among stress hormones, the master regulator of stress response in the brain, corticotropin-releasing hormone, or shortly CRH, seems to be especially important. However, how these hormones act and interact to change the placid female into a vicious defender of her brood is not yet understood beyond the point that mom's aggressive protection of children developed as a part of a more extensive plethora of maternal behaviors. The aim of these behaviors is to ensure the survival of the offspring and, consequently, the successful transfer of the genes into the next generations.

Since the dawn of humanity, aggression and love have been with us, but this does not provide excuses for behaving violently. Even though chimpanzees might wage wars just as we do, our civilized society should abhor the violence. Yes, some hormones might make us more prone to violence, but our conscious brain always has the means to overcome these primeval urges.

Love and hate, or love and aggression, might have a similar biological basis, but they are not intrinsically inseparable. Just as mothers are not aggressive per se, but only when they need to defend their nests, there are no excuses for violence and aggression. They should and could be controlled by our conscious brain. And luckily, this does seem to be happening with the progress of humankind. Even though we might think that there is too much violence in our contemporary world, historical facts show that violence was much more prevalent in the near and distant past. So perhaps, there is a hope that we can overcome our primal instincts and can grow into more peaceful societies.

# Chapter 17
# Kinky Giraffes

*With regard to large-sized vivipara, the hind only very rarely sustains the mounting of the stag to the full conclusion of the operation, and the same is the case with the cow as regards the bull, owning to the rigidity of the penis of the bull. In point of fact, the females of these animals elicit the sperm of the male in the act of withdrawing from underneath him; and, by the way, this phenomenon has been observed in the case of the stag and hind, domesticated, of course. Covering with the wolf is the same as with the dog. Cats do not copulate with a rearward presentment on the part of the female, but the male stands erect and the female puts herself underneath him; and by the way, the female cat is peculiarly lecherous, and wheedles the male on to sexual commerce, and caterwauls during the operation. Camels copulate with the female in a sitting posture, and the male straddles over and covers her, not with the hinder presentment on the female's part but like the other quadrupeds mentioned above, and they pass the whole day long in the operation; when thus engaged they retire to lonely spots, and none but their keeper dare approach them. And, be it observed, the penis of the camel is so sinewy that bow-strings are manufactured out of it. Elephants also copulate in lonely places, and especially by river-sides in their usual haunts; the female squats down, and straddles with her legs, and the male mounts and covers her. The seal covers like all opisthuretic animals, and in this species the copulation extends over a lengthened time, as is the case with the dog and bitch, and the penis in the male seal is exceptionally large.*

Aristotle: History of Animals

 
G. Majdic, *Soul Mate Biology*, Fascinating Life Sciences,
https://doi.org/10.1007/978-3-030-67212-6_17

Production of the offspring is of paramount importance for evolution. Only through our descendants our genes survive and our species thrive. To achieve the survival of the species and to transfer the genes into the next generations, animals have to mate. For animals to be as efficient as possible in the production of offspring, natural selection invented numerous ways that lead to reproduction, the ultimate goal of evolution. One of the ways is human love. In animals, the principal drive of reproduction is sexual instinct, fueled by hormones.

Sexual instincts can be divided into two domains. One is sex drive, a force that propels individuals into seeking sexual intercourse. Another is sexual behavior, triggered by a sex drive and defined as all the behaviors necessary and sufficient to ensure successful fertilization of eggs. Sexual drive is one of the strongest and oldest instincts, comparable to seeking food. Even very primitive animals such as worm *Caenorhabditis elegans* with a nervous system comprised of only a few hundred nerve cells display an intricate pattern of sexual behavior. Male worms often leave the area with plentiful food and water in search of females, suggesting that sex drive is even stronger than hunger.

The same hormones that promote and control the development of sperm (androgens) or eggs (estrogens and progesterone) also trigger and regulate sexual behavior. Testosterone is carried by the blood through the body and triggers feelings of desire and typical movements during mating. In many female animals, estradiol and progesterone initiate a typical mating posture, called lordosis.

Interestingly, almost all mammals have intercourse in similar sexual positions as described already by Aristotle, with the male straddling the female from the back. Female lordosis is a reflexive posture assumed by the females in estrus that facilitates mating in such typical position. It usually includes an arched back and lifted backside, which eases a male entrance into the vagina. However, despite universal position during sexual intercourse, diverse mating patterns have developed during evolution in mammals.

Mating can be very short or very long. In some animals like dogs, males and females are locked together for some time after mating. They often cannot separate for up to 30 min after the ejaculation. Some males just enter the vagina and quickly ejaculate. In contrast, others have to make thrusting moves before they discharge the semen into the female.

Almost every group of animals has some unique patterns of sexual behavior, although there are some general principles. Predators are much more likely to be locked together after the sexual intercourse than prey, as the latter cannot risk being immobilized for a considerable amount of time. Most prey species have quick mating systems that shorten the exposure to the predators during the intercourse. Nocturnal animals often engage in longer sex, as it is safer to engage in a distracting activity such as sex during the night. Guinea pig males ejaculate during the day after single intromission into the vagina. Nocturnal rats, on the other hand, perform multiple intromissions and thrusts before they ejaculate.

The ultimate aim of sexual engagement is to produce progeny. Males developed different strategies to ensure that they are the ones that will sire an offspring. One possibility to ensure paternity is monogamy, but this is rare in mammals. Another option is to monopolize the females by not allowing them to mate with other males. Such monopolization is seen in many animals, such as gorillas, hippopotamuses, elephant seals, and many others.

Besides social patterns, there are some tricks during sexual intercourse that can help males of some species to ensure their fatherhood. One is a copulatory lock, the physical joining of male and female for some time after sex, as is seen in dogs. Locking is caused by swelling of the penis inside the vagina and cramping of the vagina during sexual intercourse. Locking in dogs can last up to 30 min after the mating, and it ensures that no other dog will mate within the same period. This provides the time window for sperm from the first suitor to reach the egg(s). Even if a bitch mates again after the initial copulation (what they do), the sperm of the first partner will have a considerable advantage in the race to reach the eggs by the time the second partner will ejaculate. The first lover will, therefore, more likely be a father.

However, dogs usually have more than one puppy, and it is not unusual that puppies from the same litter have different fathers. This happens as puppies in dogs are not identical twins. Each pup originates from a separate egg and sperm, and not all eggs are released into the oviduct simultaneously. When in heat, the dog's ovaries release several eggs. If the bitch mates with several males within a short period, it is quite possible that sperm from different fathers fertilize eggs.

Several animals use copulatory plugs for the same purpose as dogs use the copulatory lock. After ejaculation, males of many animals close the entrance into the vagina by the copulatory plug, a gelatinous matter secreted during ejaculation, covering the opening into the vagina. Male rats and mice, as well as primates such as baboons, macaques, and chimpanzees, all use copulatory plugs. The copulatory plug has two roles: preventing the semen from leaking out of the vagina and stopping other males from mating with the same female.

Consequently, the copulatory plug is often present in promiscuous animals, while there is no need for such tactics in monogamous animals or animals where one male monopolizes multiple females. Thus, copulatory plugs are not found in humans, gorillas, or gibbons but are present in chimpanzees and other promiscuous primates. In truth, copulatory plugs do not prevent multiple matings as either male or female could readily displace the plug, but plugs delay the next coupling. This delay gives the first suitor a better chance that his sperm will first reach the eggs, the same effect as in dogs with the copulatory lock.

Although Aristotle wrote in his *History of Animals* that hedgehogs mate face to face, it is known that he erred. Both hedgehogs and porcupines mate by a male straddling a female, despite their spines, which they carefully avoid during the intercourse.

The only animals that regularly mate looking at each other are humans and our closest relatives, bonobos. There are some reports about face to face matings also in gorillas and orangutans, and perhaps lions. But in these animals, such intercourse is exceptional and not a usual mating pattern.

Despite monotony concerning the sexual position, habits before and after sexual intercourse are very varied in the animal world. Sexual intercourse is an instinct, and even in humans, it is not necessarily connected with love. Yes, sex is an integral part of love, but humans can have sex without love, at least temporarily. In this, we are similar to most other mammals, where sexual intercourse is usually not connected with emotional attachment or bonding between males and females. However, there are few exceptions when animal couples stay together after the mating, and sex in these animals is a crucial trigger to initiate the bond between males and females.

Sex is one of the most primeval instincts. Just as we need to eat food to survive as individuals, we must have sex to survive as a species. Despite being such an essential part of nature, sex is in many cultures seen as something shameful, something that must be hidden and done in privacy. While most animals mate in public spaces, our social norms dictate that sex is one of the most private things in our lives. This has been ordained through the centuries by many religions.

Some religious philosophers even tried to find parallels between animal behavior and prescribed religious norms. Saint Francis de Sales, a French clergyman from the sixteenth century, thus wrote in one of his works: "The elephant, not only the largest but the most intelligent of animals, provides us with an excellent example. It is faithful and tenderly loving to the female of its choice, mating only every third year and then for no more than five days, and so secretly as never to be seen, until, on the sixth day, it appears and goes at once to wash its whole body in the river, unwilling to return to the herd until thus purified. Such good and modest habits are an example to husband and wife."

This description of elephant behavior is, of course, not accurate, as we know today. Although female elephants usually retreat from the herd to have sex with a male, they remain in the estrous for much longer than 5 days. During this time, they often copulate with more than one bull. And they certainly do not bathe in the river after sex to purify themselves.

Some animals engage in strange, even kinky sexual habits. We have all heard about praying mantises sometimes (not always!) biting off the head of their lovers and female spiders devouring their partners. In mammals, matings are usually not such dangerous affairs, although they could be somewhat painful. Cats' penises are covered with spines. When tomcat withdraws his penis from the vagina, this causes pain to the female, often expressed with a loud scream. The reason for such painful copulation is unknown, but it might be connected with ovulation. Namely, in cats, the eggs are not released from the ovary spontaneously as in most other animals and humans but only if cats have sexual intercourse. Pain caused by the spines on the penis seems to be the trigger that causes the release of the eggs from the ovary. However, what is the evolutionary reason for such a mating pattern is not clear. Perhaps males with spines on their penises are more likely to cause ovulation and father offspring.

There are some other interesting practices found in the animal world. Many mammals rely on the pheromones to work out who is willing to mate at any particular time. Most mammals detect pheromones through the air, but not the giraffe. When a male giraffe finds a female partner that might be interested in mating, he nudges her with his snout in the genitals until she begins to urinate. The male collects discharged urine in his mouth and only then can he be sure that the female is indeed in heat and is willing to mate. Yet, even when he detects the pheromones in the female urine, she might not be ready to stand still and allow the male to mount. The male often has to follow a female for hours, sometimes even days, before she agrees to mate. Sometimes, even the tasting of the urine and persistently following a chosen sweetheart are not sufficient. If another, stronger, and more desirable mate comes along, a persistent suitor is often rejected, and the lady giraffe elopes with another partner.

Porcupines also have some kinky habits, although they prefer to keep a distance during the foreplay. Once a male porcupine finds a suitable female, he carefully guards her. He fiercely attacks any male intruders, trying to steal her heart. Often, he climbs on the tree above his sweetheart to have a good view of any potential interlopers. While he is guarding his love and waiting for her to be ready to mate, he urinates on her. It sounds like a strange habit for foreplay, but the male urine contains pheromones. These pheromones stimulate the female to enter the final phase of the estrus and arouse her to be willing to mate. As porcupine brides, unlike elephants or giraffes, remain in heat for only 12 h, they often mate with only one male, but with this one often, until the groom is completely exhausted.

If giraffes' and porcupines' foreplay seem repugnant, wait for the hippopotamuses. They do not urinate on each other but secrete the pheromones with their feces. When a male hippo detects a female in estrus, he moves closer to her. As a sign of affection, or as a tactic to win her heart, he urinates and defecates. But not only that. He moves his tail simultaneously, forming a spray of feces and urine flying around the female. Like manure spreader throws the manure all over the fields when pulled by a tractor, so does male hippo throw chips of his feces all around the female. This seems acceptable, even desirable, for the female. She is willing to accept the male's advances only after such a strange display of love.

Once the couple agrees on mating, they move to the water. The male then climbs on the top of the lady to perform mating similarly to other animals. However, as male hippos are massive, females are entirely submerged under the water during intercourse. The mating could last for up to 30 min, and females are underwater most of the time. Luckily, females can hold the breath for a long time as if a lady reaches out for some fresh air, the male, very ungentlemanly, often snaps at the female head with his big and scary mouths and forces her back underwater.

There are other, perhaps not so kinky, nevertheless interesting and somewhat strange mating habits present in other animals. In most mammals, mating is a quick affair that usually lasts a few seconds to a few minutes. In some animals, it takes much longer. Kangaroos' sex lasts up to 50 min, and boars often have an orgasm lasting 30 min or more. During this time, boars also ejaculate a huge amount of semen, up to half a liter of semen during single mating.

However, when it comes to devotion, stamina, and self-sacrifice, the strong contenders for the champion title are small, mouse-like marsupials from Australia, called *Antechinus*. Although virgins until the first sexual encounter, male *Antechinus* cannot be accused of being half-hearted when it comes to sex. They reach sexual maturity just shortly before their first birthday, right when the mating season starts.

Males mate incessantly with any willing female for 2 or 3 weeks. One sexual encounter can last for up to 14 h, and males do little else during this sexual spree. Such exhausting behavior has consequences on the body, and male *Antechinus* bodies literally disintegrate during this lustful period. Their bodies are filled with testosterone, and male *Antechinus* care about nothing but sex. Subsequently, their immune system collapses. Their fur falls off. They bleed internally, and shortly before their first birthday, all the males die of exhaustion. They sex themselves to death and are thus an example of the highest devotion to the survival of their species.

Why has such behavior evolved is unclear. The death of the males could provide more resources for pregnant females and later for their offspring, as males would not consume scarce food. However, this behavior could also be just some quirk of *Antechinus*'s evolutionary history. Another option is that this is just a sperm war, progressed to the extreme, with males competing with each other who will mate more often and with more females.

In promiscuous mammals, males try to mate with as many females as possible to sire the most offspring. In gorillas, single alpha male wards off all other males from mating, and he sires all the offspring. However, promiscuous chimpanzee females often mate with many males during a short period. The male who impregnates the female with the most sperm has the highest chances statistically to become the father, and this competition between sperm is sometimes called the sperm wars.

*Antechinus* seems to bring this to the extreme. Males prepare for almost a year for the ultimate race between the males—who will be more successful in mating and mate with most females. For this, males are giving the ultimate sacrifice—their life—to increase their chances to transfer their genes into the next generation. Such behavior, as observed in the *Antechinus*, is rare among mammals, which are usually more restrained when it comes to sex. Yet, *Antechinus*'s behavior clearly shows how powerful is the sex drive.

Testosterone primarily stimulates sex drive in males. Castration of animals usually eliminates their interest in the opposite sex, and giving them back testosterone reinstates their sex drive, although sometimes not immediately. The sex drive will not return at once if there is a lag between the castration and reintroducing testosterone. Once without testosterone, the body and brain get insensitive to testosterone. When testosterone returns into the body, the organism has to develop sensitivity again. It might take several weeks before males are ready to mate again, but eventually, they will become horny.

Testosterone is essential for both sex drive and sexual behavior, but different parts of the nervous system regulate these two instincts. The medial preoptic area, the same brain area that is important for maternal behavior, has a pivotal role in regulating sexual behavior. Male rats with the damaged medial preoptic area do not mount females. They do not copulate, but they are still interested in females. They will run through the labyrinths and press levers to get access to females, but once male and female are together, males simply do not know what to do and why they are there. The amygdala, on the other hand, is the central region responsible for the sex drive. Pheromonal and other signals derived from the female are relayed to the amygdala. The amygdala decodes these signals and transmits the messages to the medial preoptic area. Under the influence of testosterone, the medial preoptic area triggers the copulatory behavior, and the male is ready to mate.

In animals, it seems that there are two crucial components to the proper male sexual behavior. One is sensitization of the nervous system by testosterone, and another is the activation of the amygdala by external stimuli. Testosterone acts at multiple levels. It is essential for the erection or penile reflex in animals and humans by sensitizing an area in the spinal cord that triggers erection, a purely autonomous reflex. For complete sexual behavior, testosterone also has to sensitize the medial preoptic area, which seems to be the control module for integrating various signals connected to the mating. The medial preoptic area must be activated by testosterone for the proper execution of the copulatory behavior. However, copulatory behavior does not begin until signals about the presence of external stimuli (willing females) arrive from the amygdala. Only when the medial preoptic area integrates hormonal signals (testosterone) and external signals from the amygdala, the complete sexual behavior from motivation to copulation is executed.

Similar to testosterone in males, female sex hormones estradiol and progesterone are essential for female sexual behavior. At the beginning of the estrous or menstrual cycle, the production of estrogens steadily increases in the ovary. These hormones sensitize brain areas responsible for female sexual behavior. When the egg is released from the ovary at the time of ovulation, the ovary abruptly reduces its production of estrogens. It starts to secrete large amounts of progesterone. Progesterone has essential roles in sustaining early pregnancy, but at the time of ovulation, it also triggers the capacity for female sexual behavior.

This is true in most animals, although in some like hamsters and voles, it seems that elevated concentrations of estradiol are sufficient to prepare the female for mating, without a need for progesterone. However, like in males, the hormones are not enough to induce the female's copulatory behavior. Estrogens and progesterone are necessary to prepare the brain for regulating copulatory behavior, but external stimuli are needed before the female shows typical sexual behavior. Unlike in males, these external stimuli are primarily tactile and not visual or pheromonal. Males, ready to mate, approach the female in estrous in various ways specific to different species. Tomcats bite the female at the scruff of the neck, male rats and many other males mount the female from the back, and stallions often bite a mare on the neck. Males often grab females on the flanks or put pressure around the perineum or at the base of the tail. Skin receptors transmit all these various tactile stimulations to the female's brain where they trigger female copulatory behavior if estrogens and progesterone previously sensitized the brain.

In humans, sexual drive is as strong instinct as in animals, but copulatory behavior is less influenced by hormones than in animals. In many animals, sex-related hormones directly trigger specific behavioral patterns, including typical copulatory positions. Female rat or mouse given estradiol and progesterone will instinctively assume standard lordosis position when in the presence of a male, following tactile stimulation of the skin. Male rat, enthused by olfactory or visual stimuli, will mount and try to intromit into the vagina. It will follow an archetypal sequence of movements, regulated by the brain previously sensitized by testosterone. Similarly, the vast majority of primates follow particular mating patterns controlled by special brain circuits. Interestingly, almost all mammals, including primates, display somewhat similar copulatory behavior with a rear entrance.

In contrast, humans have sexual intercourse in various positions and follow different behavioral patterns. This suggests that sexual behavior is not as strictly hormonally regulated and is not as reflexive an undertaking in humans as it is in animals. In contrast to animals, we also do not have sex only during female ovulation, so sexual desire in women is not critically dependent on the estrous hormones. Therefore, it seems that sex in humans has additional dimensions beyond reproduction.

Sex in our closest relatives, bonobos, is an important social activity and does not have a strict reproductive role. Similarly, sex in humans has a broader role beyond reproduction. It is an essential aspect of love, and love usually does not exist without sex. Whether sex lives without love is not clear. Some claim that casual sex does not exist, as hormones released during sex and especially orgasm, such as vasopressin and oxytocin, trigger feelings of attachment. These hormones could, therefore, cause trouble. Even when we enter a no-obligations sexual relationship, we might end up being in love. Oxytocin and vasopressin could unconsciously force parts of our brain to become enamored and attached to our sexual partner. So instead of some fun, we might end up in a serious relationship.

Although sex is undoubtedly an integral part of love, sexual desire or lust often seems to be separated from true love and happily ever after. But, just as one kiss persuaded Snow White to fall for a handsome prince, a casual sexual affair might turn our lives upside down, as we might fall in love unexpectedly.

# Chapter 18
# Dik-Dik: A Faithful Dwarf Antelope

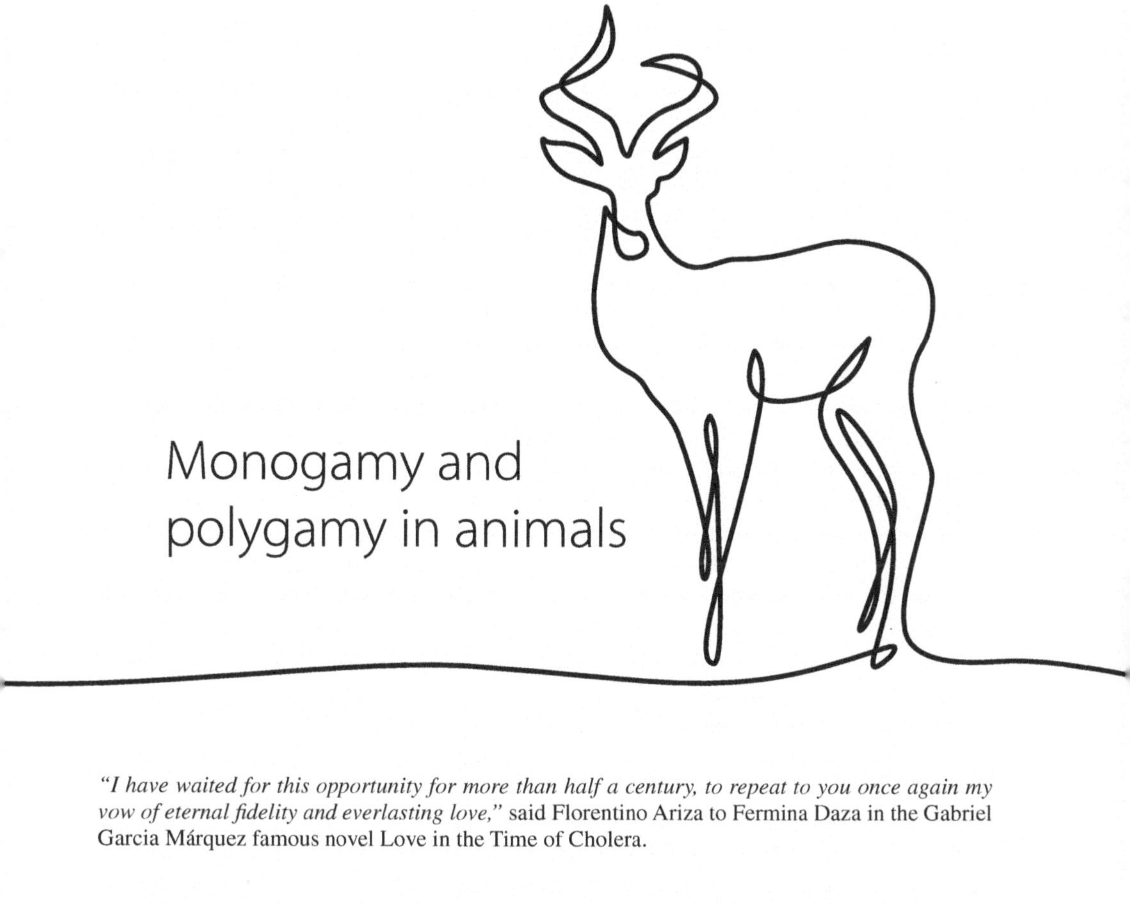


*"I have waited for this opportunity for more than half a century, to repeat to you once again my vow of eternal fidelity and everlasting love,"* said Florentino Ariza to Fermina Daza in the Gabriel Garcia Márquez famous novel Love in the Time of Cholera.

Yet, this was an unfulfilled love. It is not a description of a long-lasting monogamous relationship, usually associated with love in contemporary western societies. Only about 15% of human societies practice strict monogamous relationships. But even within the cultures that are agreeable to polygynous (one man having several wives) or polyandrous (one woman having more husbands) relationships, most individuals engage in monogamous relationships.



G. Majdic, *Soul Mate Biology*, Fascinating Life Sciences,
https://doi.org/10.1007/978-3-030-67212-6_18

This pattern suggests that humans are primarily socially monogamous. However, like birds, it is not uncommon for humans to be sexually polygamous even when they are cohabiting in a socially monogamous bond. We, as a species, are also practicing serial monogamy. After a break-up, divorce, or the partner's death, we usually find a new partner. Many cultures, not only Judaism and Christianity, cherished monogamy throughout history, and till death do us part is present in numerous societies. Yet, we do not know if these habits have cultural or biological origins.

Already in the Hammurabi code, the oldest known legal system, monogamy was the preferred type of cohabitation. However, if a woman did not bear a child, a man was allowed to take another wife. But there was a catch. The Hammurabi code also had a special provision for a childless wife. She was allowed to give his man a slave or a servant girl as a gift so the man could have fathered children with this slave. In such a case, a man was not allowed to take another wife. So, an assurance was built into the code to preserve the outwardly monogamous bond between husband and wife.

Monogamous behavior makes us somewhat an exception in the realm of mammals. We do not know the exact number of mammals practicing monogamous relationships, but the estimates are that only between 3% and 9% of all mammals are strictly monogamous. This percentage is meager compared to birds, where around 90% of all birds practice social monogamy, one male and one female living together. However, these partners do not necessarily mate exclusively with each other, something not unheard of also in the *Homo sapiens*. Birds are often socially monogamous, but they are not necessary sexually monogamous. These feathery monogamous relationships are also short-lived, as many birds change their partners every breeding season. In contrast to birds, monogamous behavior is not a norm in mammals, and it is still unknown why it has developed in some animals. Numerous hypotheses were put forward in recent years and decades, but it is difficult to prove any of them.

The main drive of evolution is reproduction and spreading genes of each individual into as many offspring of the next generation as possible. The variety of novel genetic combinations is beneficial for the survival of the species in ever-changing environments. From a purely evolutionary perspective, it would be beneficial for both sexes to seek as many partners as possible to produce large numbers of novel genetic combinations in their descendants.

Unlike fishes and reptiles, who sometimes deposit an enormous amount of eggs into the environment, mammalian females are severely limited in the number of offspring they can produce. Only a few eggs develop in each sexual cycle in mammalian ovaries, restricting the number of babies one female can have. Mammalian babies grow inside their mothers, and this presents another limitation. The uterus and belly can hold only a limited number of babies. Expectant mothers sacrifice their own resources to provide nutrients for the developing fetus, and these resources are always limited.

Furthermore, mammalian babies need maternal care after birth. Within the first weeks or months of their lives, infants cannot survive alone. They need the food in the form of milk, and mothers often have to warm them and protect them from all possible dangers, including predators. Thus, mammalian mothers are very limited regarding the number of children they can raise.

On the other hand, males do not take part in the pregnancy. They do not have to sacrifice their resources to provide nutrients to the developing baby in a womb or to produce milk for the babies. Male testes continuously produce enormous numbers of sperm that could be easily wasted. So vast are the numbers of sperm that when scientists initially observed human sperm under the microscope in the late seventeenth century, heated discussions began. Many scholars claimed that sperm have nothing to do with the creation of new life but are some kind of parasites. This notion derived from the idea that God would not produce so many new living beings that are wasted with every discharge of the semen.

With a vast number of sperm, males have a much better opportunity to spread their genes widely. Males can mate with multiple females. They can produce many descendants carrying their genes, thus ensuring that their genes are preserved in future generations more efficiently than females. Considering all these, there is seemingly a conflict in the interests of mammalian males and females.

Moms must provide dedicated maternal care if they want to ensure the survival of their genes in the next generation. Fathers could just impregnate the female, bugger off, and be hopeful that some of the many females they had sex with will successfully raise their descendants. Therefore, mothers prefer a close bond with an affectionate father of their children. They must find a father who will help them care for the progeny.

For fathers, a better strategy is to produce a large number of children what should ensure that at least some of these descendants will carry their genes into the next generations. This apparent conflict situation led to the appearance of various mating strategies in mammals. It seems that male strategy is more successful in the long term, as most mammals are polygamous. Most mammalian males try to mate with multiple females to propagate their genes and produce a large number of novel genetic combinations.

Guys can use different methods to achieve the goal of spreading their genes widely. The first and the most obvious is to win the hearts of as many ladies as possible and beat other fellas trying to do the same. For this purpose, many males developed unique ornaments like antlers in deer, manes in lions, horns in antelopes, colorful face in mandrills, or larger and more muscular body size in many mammals.

All these serve two purposes—winning the female's heart and beating an opponent in close combat for a sweetheart. In the vast majority of animals, there is fierce competition between males who will impregnate more females and thus more efficiently spread his genes. The most successful male is usually the deer with the most massive antlers, ibex with the most impressive horns, a lion with the thickest mane, and boar with the largest canines. Consequently, natural selection led to the development of sometimes enormous ornaments or differences in animals' body size.

However, as males compete for the same females, they often find a female too late, especially if animals are not living together in a group. Wandering lion often finds a lioness already with cubs, sired by his competitor. Young and strong gray langur could fight his way to the top position in the troop, to find out that his recently defeated predecessors fathered all the babies. In many animals, males are taking very drastic and cruel measures to counteract this. They attack and kill the babies. This infanticide prevents the genes of the male's opponent from being transferred into the next generation and, in many species, induces heat in the female.

Most mammalian mothers are temporarily infertile while breastfeeding their babies (although this mechanism is not entirely faultless, so young mothers should not rely on this for contraception). If a male kills all the babies, the mother soon stops lactating and enters the heat. Such a female is willing to mate with the new male, even though the fella slaughtered her babies just a few days earlier.

To prevent the killing of the offspring, females have two choices. One is to mate with numerous males at the same time. Males will not know who the babies' father is and will not risk killing their progeny. Indeed, infanticidal behavior is rare, although not absent, in highly promiscuous animals such as chimpanzees. Another option to preserve the babies is to convince the father to help protecting the young. One of the hypotheses, why has monogamous behavior developed, derives directly from this.

Every male is interested in preserving his genes into the next generation. If a male lion kills another father's babies, genes of the unfortunate father are lost. Such dad was unsuccessful in fulfilling his primary evolutionary task, transferring his genes into the next generation. Therefore, it is beneficial for a father to stay with the mother throughout the pregnancy and during the infancy of their babies. Mother and father together will defend the infants much easier and better, and this certainly seems a logical evolutionary path toward monogamy.

In some cases, being faithful to each other is essential for the survival of the infants. Such a strategy is especially important in animals that live in an environment with a scarce supply of nutrients. Mimic poison frog, living in South America, is one of only a few amphibia with both parents caring for their young in a seemingly monogamous relationship. Tadpoles hatch in tiny pools of water with a minimal supply of nutrients. To keep the tadpoles alive so they can develop into adult frogs, both parents have to feed them. Tadpoles of mimic poison frog, unlike tadpoles from most other frogs, cannot survive alone as their environment does not provide enough food.

Lack of food has most likely triggered the appearance of the monogamous partnership in these frogs. In beavers, monogamy probably arose from opportunistic behavior. Beavers live in large nests made from tree branches, built in the middle of the rivers. They have to repair their nests continually. Mending of the nests presents a considerable workload in addition to caring for their young, which is, as most of us know, also a very demanding job. So, it does make sense for the mister beaver to help his lady take care of the house and children. Children might be otherwise neglected and would have a lower chance of survival.

Another hypothesis about the development of monogamous behavior is more simplistic. Some evolutionary studies suggest that early mammals lived mostly solitary lives. Although many animals today live together in large herds with sometimes enormous number of females, many animals are still solitary. Blokes have to wander around, sometimes covering vast distances before they could find a suitable mate.

Such wanderings mean spending a lot of energy while searching for females. During roaming, you expose yourself to predators. And finally, you are always under the pressure that another fella will find the lady before you. Therefore, it makes sense to stay with your missis once you found her and seduce her into the mating.

One example of monogamous animals is the small cute African antelope dik-dik, with large beautiful eyes and charming eyelashes. Unlike most antelopes, dik-diks are not very social animals and do not like to spend their time in large herds. For many antelopes, protection from predators is simply a large number of animals. They are willing to sacrifice a few members of the pack for the greater good of the many.

Dik-diks and their relatives klipspringers, duikers, grysboks, and other dwarf antelopes adopted a different strategy. They live alone or in tiny groups. They rely on their small size and ability to hide in a tall savannah grass for protection from the predators. Lonely dik-dik females live in a defined territory. As these territories are sometimes quite large, males could have difficulties finding a single mate, let alone multiple ones. Therefore, it makes perfect sense for a male dik-dik to stay with his soulmate once he finds her, as he could never know when he will have a chance to encounter another love. Dik-diks are not exemplary fathers. They rarely help the female with chores, so it seems that sticking with their soulmates is not based on love. Male monogamy is more likely only due to the opportunistic reasons that another female might be difficult to find.

Some have questioned this and have shown that dik-diks are, despite being poor fathers, faithful mates. They would not leave their partners even if they have an opportunity to mate with another bachelor female. So, after all, perhaps there is something akin to long-lasting love in the dik-diks' partnership.

For similar reasons, monogamous behavior probably developed in prairie voles and even in intestinal worms called *Schistosoma mansoni*. These worms live inside our intestines. As we can imagine, the insides of our guts are not the most romantic place to find a soulmate. So, it is quite understandable that the female worm clings to the first mate she encounters and stays with him for life. She clings to the partner literally, as once she finds a guy, the female nests inside the male in an organ called gynaecophoric canal. She spends the rest of her life inside her mate. After all, there is probably a slim chance to find another, more agreeable mate inside the dark, smelly human intestine if you let one partner slip through your fingers, or to be precise, out of the gynaecophoric canal.

Whether mammals are monogamous or polygamous is often possible to predict from the characteristic of their bodies. In animals with fierce competition between blokes to win the soulmate's love, males and females are usually very different in appearance. They might be very different in body size like male and female gorillas. Males could carry formidable weapons such as canine teeth in boars or possess ornaments such as antlers in deer and mane in lions. Both antlers and manes serve as attacking or defensive weapons, especially in fights between males competing for females. However, ladies also carefully observe the size and appearance of ornaments or muscles on the bones when choosing the mate. Willing females usually want the shapeliest male to be the father of their children. It is likely that picky ladies drive the evolution of beautiful ornaments or sometimes ridiculously large weapons in males.

One might think that birds are an exception to this rule about flashy males, as a vast majority of birds are monogamous. Yet, male birds carry some of the most elaborate and colorful ornaments in the animal world. This is not a paradox, as the majority of birds only mate for one breeding season. Males have to find a new soulmate each season. Hence, their beautiful plumage and wonderful singing are essential in the fierce competition between guys who need to find a new partner every spring.

However, let's look at thoroughly monogamous birds. We can see that in swans, bald eagles, barn owls, penguins, and others that mate for life, sex differences are not prominent as they do not have to find soulmates again and again. In mammals, we can see a similar pattern. In truly monogamous animals, there is, as a rule, a small difference between males and females. Everybody can easily say who is male and who is female when looking at a deer with antlers and a male and female gorilla, orangutan, or wild boar. But we would struggle to find out who is a girl and who is a boy in wolves, dik-diks, gibbons, or beavers, without checking their genitals. However, checking the genitals might be a tricky business, especially with animals like wolves, as both sexes have very sharp teeth.

Another natural sign of whether animal is promiscuous or not is the size of the balls, or, more scientifically, testes. In animals where a female has sex with only one male, testes are, in general, smaller. If males and females live in monogamous relationships, or if one male controls the group of females, testes produce fewer sperm as there is no competition between males who will impregnate the female. In promiscuous animals, testes are, as a general rule, large. Females in promiscuous animals do not choose only one partner but often mate with more than one fella during one estrous cycle. In such a case, it is vital to produce more semen and a higher number of sperm. More sperm give the male a better chance of being successful at fertilizing eggs inside the female. Just by chance, the higher the number of sperm, the better the odds one of yours will reach the egg.

One example of such animals is our close relatives chimpanzees. Chimpanzees live a very promiscuous life, with matings going on between various members of the troops. Females within the troop often mate with two or more males within a short period of time. In a fierce competition whose sperm will win the race and fertilize the egg, chimpanzee testes grew to a formidable size during evolution. Testes are almost as big as their brains, which are not the smallest in the animal kingdom.

In contrast to chimpanzees, male gorillas are pretty poorly endowed in that department. In gorillas, one patriarch, the largest male in the troop, the silverback, has sex with all the females. Subordinate males do not even dare to approach the females in fear of the leader's reappraisals. Therefore, there is no need for a large number of sperm in the ejaculate. Once a silverback climbs to the top of the corporate ladder, he can be pretty sure only his sperm will be inside the females, as long as he is the boss.

As we can see, there are specific body characteristics in mammals from which we can predict reasonably accurately who will be living happily ever after and who is always seeking greener pastures. So, what does all this mean for the *Homo sapiens*? Are we monogamous or polygamous? Is being unfaithful to our partner wrong, or is it something embedded into our genes and brains and we cannot control it?

Let's start from the end—we are a conscious species with an extensive brain area devoted to thinking, judging, and reasoning. Almost every decision we make is our conscious decision, except instinctive reactions when we are in danger. So, whether to cheat or not to cheat on our partner is our choice, and we cannot blame biology for any such decision we make with our conscious brains. Yes, love can overwhelm our senses and parts of our brain, but only for a short time and never so much that it would completely switch off our conscious brain called the prefrontal cortex.

Culturally, it seems we are mostly monogamous. In most human societies, even in those that do not sanction polygamy, people most often live in monogamous relationships. But has this monogamy developed together with the civilization? Or was it deeply embedded into our brains before the dawn of sapient human beings? If we look at our natural characteristics, men do not carry any particular ornaments provided by nature. Yes, men have beards and Adam's apples, but these can hardly compete in size and impressiveness with deer's or moose's antlers.

Yes, men can buy flashy sports cars and invite girls to expensive parties, but these are not biological characteristics. If we look at the body size, men are slightly, but only marginally, larger than women. This is in stark contrast to some of our relatives, promiscuous apes like gorilla and orangutan. In gorillas and orangutans, there is no monogamy. Gorillas live in closely attached family groups, led by an alpha male, a silverback patriarch. He monopolizes all the females and prevents all other males from approaching them. The leader of the troop is always the largest male. Fierce competition and fighting are going on, and quite logically, the largest and the strongest male wins and grabs the price—all the females.

Orangutans, especially adult males, live a solitary life with couples meeting only to reproduce. Males, however, do occasionally fight fiercely to win the heart of the loved one. As an evolutionary consequence of fighting, males who need to wrestle with other males are much larger than females, which is true in both orangutans and gorillas. In gorillas, silverback male is, on average, 1.5 times larger than the average female. In orangutans, dominant males are often twice the size of the females.

Chimpanzees and bonobos are promiscuous animals, living in large troops, with males being more tolerant of each other than gorillas. Sex is happening between various troop members, and girls often have intercourse with more than one guy within a short period. In these apes, the difference in body size between sexes still exists but is much smaller. Average chimp is about 1.3 times larger than female and bonobo 1.2 times. In contrast, another ape species, gibbons, live in very harmonious monogamous relationships. Males do not fight for females, and there is no need for boys to grow larger and stronger. Indeed, the sex difference in gibbons' size is almost non-existent. On average, an adult male gibbon is only 1.02 times larger than an adult female.

In humans, this difference is 1.1, meaning that men are, on average, 10% larger than women. This difference places us somewhere between promiscuous bonobos and truly faithful gibbons. It seems that, in principle, biologically, we are more monogamous than promiscuous animals.

Quite similar is the situation when we look at the testis size relative to body weight. Relative testis size is a more appropriate measure than an absolute testis size as absolute size is undoubtedly larger in bigger animals and is thus not informative. Chimpanzees and bonobos have very large testes for their body size, presumably due to their mating habits. As females mate with different males within a short time span, they often have sperm from more than one lover inside her genitals. There is fierce competition between males whose sperm will reach the eggs. Hence chimp fellas evolved ever-larger testes, which are now formidable in size, larger than the chicken egg.

In gorillas and orangutans, testes are much smaller in relation to their body weight, as females in both apes mate with only one male at a time. Consequently, there is no sperm competition within the females. Similarly, men are not as well-endowed as chimpanzees. The average human testis is only the size of a walnut, much smaller than the chicken egg. Yet, relative testes' size of the average human is slightly larger than gorillas', and some suggest that we are therefore not genuinely monogamous species. However, looking at the relative testis size, we are much closer to gorilla and orangutan than to chimpanzee or bonobo. So, this suggests that biologically we have developed to be with one partner, rather than being a promiscuous species.

Another suggestion that monogamy is part of our nature is concealed ovulation. In most animals, females are only ready to have sexual intercourse around ovulation, when they are in heat. On the other hand, humans have sex often and are not limited to the time of ovulation. If we would be restricted to the time of ovulation, this could present a significant problem. Unlike many female animals, women do not show visible signs when they are ovulating and are most fertile. Some suggest that this is another sign of monogamy, as in females with concealed ovulation, a male should stick around all the time.

If female signals she is in heat by releasing large clouds of pheromones into the air, marking the path to her burrow with mating pheromones, or showing off with the red swollen genitals, a male can approach such a female at the appropriate time. He can mate with her and leave immediately. Male can be assured he was impregnating the female at the right time, the most suitable time for fertilization. However, if the female has concealed ovulation, the guy cannot know when the girl is fertile. In such a case, it is better to stick around the sweetheart, have sex as often as possible, and prevent other males from having sex with your loved one. Only this could ensure you will be the father of her babies.

Nevertheless, some argue that we are biologically promiscuous, and monogamy has developed only with civilization for some practical reasons. One such reason would be to keep bachelor males in rein. Young males with testosterone flowing through their bodies can be difficult to control. It is sensible to tie them down in a relationship with a woman. In a society where powerful men take many wives, many blokes would remain bachelors. Instead of using their energy toward fulfilling their sexual instincts and taking care of wives, they could become violent, and they might present a danger for the leading classes.

Another plausible reason for culturally promoted monogamy is the prevention of sexually transmitted diseases. Such diseases have played a significant role throughout human history and have decimated some populations at certain times. To prevent the spreading of sexually transmitted infections, some societies might have introduced enforced monogamy to preserve the population healthy.

When did we become monogamous? There is no exact answer to this question. Some argue with the dawn of civilization, and others that biological monogamy developed much earlier during evolution. About 4.4 million years ago, our ancestors, *Ardipithecus ramidus*, roamed across the African savannahs. This might have been our first forefather walking exclusively on two legs. Walking on two legs freed hands to do various things, perhaps bringing food and taking care of the loved one. Some evidence also suggests that differences between sexes diminished around that time, with males and females becoming more alike than were our earlier ancestors. These two facts might indicate that humans started to develop into monogamous species when *Ardipithecus ramidus* roamed in Africa.

Another point of view is that monogamy was established with the human heads' enlargement due to bigger brains, causing the birth of less-developed human babies. Hence, this prolonged the time after birth when children need care, protection, and supply of high-quality food. This happened between 1.5 and 2 million years ago, at approximately the same time when our ancestors became more proficient hunters and tamed the use of fire. These inventions all helped to provide better food needed for the growth of the larger brains. While children became more dependent on their mothers, mothers had less time to forage for food, which might have led to the establishment of pair bonding in our ancestors.

There is no proper answer to whether we are biologically monogamous or polygamous. Indeed, we cherish strong bonds between man and woman in almost all human societies. However, the occasional extramarital affairs were part of our culture since ancient times. Otherwise, there would be no need for the sixth commandment, "Thou shall not commit adultery." But whether adultery is written in our genes and embedded in our brains, we do not know, and it is unlikely we will ever have a definite answer.

We are mostly committed to our partners, but we are also serially monogamous. After breaking up or the death of a partner, we will usually find another partner. Statistically, the riskiest period for a divorce is about 5–7 years after the marriage, just as in the movie The Seven Year Itch. Some scientists suggest that our natural mating system is serial monogamy with changing our partners every 4–5 years. Four years is a period during which a human child in prehistoric societies would be large and robust enough to survive without parental care, especially without the need for a father. Four years is also an average period between women having consequent children in some hunting and gathering societies. Some use this as an argument that we are biologically serially monogamous.

However, we can also have a look from the other side. Many relationships might break up after 5 or 7 years of togetherness. But there are many more partners that stay together for a very long time, often until the partner's death. This faithfulness, in contrary to the 7-year itch, suggests that we are truly monogamous species. In the end, it is up to us. Our conscious brain has to decide whether we want to stay in a long-lasting, stable relationship and find fulfilling love in such a partnership. Another option is to fly from flower to flower without permanently settling down, at the risk of finding ourselves lonely and deserted at some point in our lives.

# Chapter 19
# Love in the World of Voles

## Why are some animals monogamous?

*Love looks not with the eyes, but with the mind, and therefore is winged Cupid painted blind.*

William Shakespeare: A Midsummer Night's Dream

We can fall in love slowly, over time, as we get accustomed to somebody, or love can strike us suddenly, like lightning. Some can fall in love unexpectedly, like Julia Roberts in Nothing Hill movie. Love can hit us at first sight, like in Sleepless in Seattle, or love can develop over time, like between Harry and Sally in When Harry Met Sally. Some people love for a very long time like Forrest Gump loved Jenny, while other will charm and seduce new partners on every occasion like James Bond usually does. Love is often complicated as the love between Scarlett O'Hara and Rhett Butler in Gone with the Wind. In movies and in real life, love is often complicated.

Although we know that we love with our minds and not our hearts, it is difficult to understand what is happening inside our brain when we are in love. We do not have the tools to examine our minds. We cannot precisely see what excites our brain cells and which brain cells talk to each other when we fall madly in love. We do not know what changes in our brain when we are attached to the person we love. We are also rather an exception in the mammalian world, as less than 10% of all mammals pair for life and spend their time in long-lasting monogamous relationships.



G. Majdic, *Soul Mate Biology*, Fascinating Life Sciences,
https://doi.org/10.1007/978-3-030-67212-6_19

Animals that scientists most often study in the laboratories, and hence we know the most about their brains, such as rats and mice, are not monogamous. We cannot learn much from them about what processes in the brain instigate a couple to stay together for a long time. We cannot study monogamy in animals that are not monogamous. We cannot learn much about the attachment between boys and girls in animals, where such attachment does not exist.

Some mammals are monogamous, though, and among these few are some voles. Voles are small, mouse-like furry rodents that live in underground burrows. They have thick gray fur but lack a long naked tail, typical for mice and rats. Voles are favored snacks for weasels, hawks, and many other predators. Therefore, the life span of voles is short. Still, some voles spend their whole life, albeit brief, together in a seemingly harmonious relationship.

Prairie vole males and females, living in the prairies in the Midwestern United States, remain together happily ever after once they have sex for the first time. Partners share parental responsibilities. They protect each other and their families. Male and female take care of the children. If one partner dies, a widow or a widower will rarely enter a new relationship. Indeed, the widowing period in voles is short due to their short life span. Yet, when prairie voles lose their partner, voles show something akin to grief and sadness.

But not all voles are made equal. Meadow voles and montane voles, in many respects very similar to the prairie voles, are both promiscuous. While prairie vole males find a partner, have sex with the lady, and stay with her for the rest of their lives, meadow vole and montane vole males wander around during the mating season to find and mate with as many gals as possible. After sex, meadow or montane vole males never look over their shoulders. They never give any toss about how the lonely mother will bring up their children.

Due to their behavior, voles have attracted the attention of scientists. Voles were partaking in numerous studies of what is happening inside the brain when animals bond together for life as humans do. Although we do not have any proof that the same processes occur in our brains when we are attached to our lifelong partners, we can assume that some of the molecular changes happening in the vole's brains might also occur in our brains when we are bonding with our partners.

An essential ingredient for the prairie voles to begin a relationship is having sex. When a vole boy and girl are placed together and socialize, but do not have sex, they do not develop a special, lasting relationship. If a couple that socialized for a few hours but didn't have sex is separated and later rejoined, the female does not show any special preference for the partner with whom she was previously acquainted. If given a choice of a formerly known male and completely new male, she does not show any fondness for the old acquaintance.

However, things change if the couple has sex during their first date. In such a situation, if later given a choice between two handsome blokes, females always choose to spend the time with their partners with whom they previously shared sexual intimacy. Sometimes, if a bloke and gal are together for a long time, they would develop mutual attachment without having sex. But as a general rule, the attachment and preference do not happen unless girls had sexual intercourse with blokes.

Oxytocin is a hormone that triggers and facilitates the attachment and bonding between mother and child in various animals. So, when scientists started to look at what is regulating the attachment between partners in voles, following oxytocin was the obvious tempting path to survey. Indeed, this proved to be the right track. It is now well established that oxytocin plays a major role in the attachment and bonding between male and female voles. If we block the action of oxytocin in the prairie vole female brain during sexual intercourse, such a female does not develop a lasting relationship with her partner. Conversely, if oxytocin is injected into the brain of a female who is socializing but not having sex with a male, the female will become attached to this particular male without having sex with him. It seems that mating triggers the release of oxytocin in the female vole brain and oxytocin somehow initiates the happily ever after.

When scientists did similar experiments with male voles, they got mixed results. Some studies reported that oxytocin also triggered the attachment to a particular gal, but others could not confirm that. The role of oxytocin in males is therefore not yet clear. But males have large amounts of another, very similar hormone, vasopressin, in their brain.

Vasopressin is, like oxytocin, produced in the hypothalamus and is structurally very similar to oxytocin. Only two out of nine amino acids, building blocks of proteins, are different between both, and vasopressin also acts in the brain. More vasopressin is present in the various parts of the male brain than in the female brain in numerous animals. In addition to the hypothalamus, other brain regions such as the amygdala also produce vasopressin. Like oxytocin in female voles, if vasopressin is injected into the male vole brain when a male is mingling with a female, this male becomes committed to the female without having sex with her. Likewise, if the action of vasopressin in the male brain is blocked during sex, males do not develop feelings of affection toward a particular girl. They wander away immediately after sex, always seeking the greener pastures.

It seems that oxytocin in ladies and vasopressin in gentlemen are the two hormones responsible for everlasting love, at least in voles. Whether the prince's kiss triggered the release of oxytocin in Snow White's hypothalamus, or whether dancing with Cinderella released the vasopressin in the prince's brain, we do not know. However, it seems that both oxytocin and vasopressin have some roles also in humans.

Oxytocin is released during orgasm in both men and women. This release is likely connected with the pleasant feelings of orgasm, although oxytocin also has more mechanistic roles. In men, oxytocin triggers the spasms of tubes around the testes, needed for the ejection of sperm. In women, oxytocin released during orgasm has a similar role as in childbirth. It triggers the contractions of the uterus and oviducts, squeezing the sperm toward the ovary. These contractions help the sperm to reach their destination, an egg released from the ovary. What is the role of vasopressin is not yet known, but some studies showed that vasopressin is released in men's brains when men are sexually aroused.

But is it just the release of the oxytocin or vasopressin during the sexual intercourse that triggers the attachment and bonding in voles? To answer this question, scientists looked at the brain of another vole species, the polygamous montane vole. Interestingly, when polygamous montane voles have sex, the surge of oxytocin in the female brain and vasopressin in the male brain also happens, quite indistinguishable from the prairie vole brains. But unlike in prairie voles, injection of vasopressin into the brain of montane vole males during socializing does not turn them into faithful partners. So, it seems it is not the release of these two hormones that make someone monogamous and someone always searching for new adventures.

What is different between prairie and montane voles is how oxytocin and vasopressin act in the brain. All hormones and neurotransmitters act through the special molecules called hormone receptors. These receptors are a bit like doorbells on the cells. If a hormone wants to kick cells into some action, it must first find and then bind to its receptor. Receptors for most hormones are on the outside of the cell membrane, so hormones that are transported by the blood can easily find them and attach to them. Binding of the hormone sends signals inside the cell, just like the electric doorbell sends an electrical message inside our house. And just like the ringing of the doorbell tell us that somebody pressed the doorbell button outside of our home, cells recognize that hormone has attached to its receptor. The same person could trigger the doorbell in different houses, but doorbells sound different, as many different types exist. Similarly, the same hormone can have receptors on various cells, but cells in different parts of the brain respond differently, as their functions are different.

The difference between which brain cells are responsive to the oxytocin and vasopressin seems to be the underlying reason why prairie voles are monogamous and montane voles are polygamous. Receptors for both oxytocin and vasopressin are present in very different parts of the brain in these two vole species. Therefore, after brain cells release oxytocin and vasopressin, these two molecules activate different brain regions in monogamous and polygamous voles.

One of the most critical differences is in the presence of oxytocin receptors in the area called nucleus accumbens and for vasopressin in the area named ventral pallidum. Both these regions are part of the reward pathway, a feel-good brain circuit. In monogamous prairie voles, but not in the polygamous meadow or montane voles, numerous oxytocin receptors are present in the nucleus accumbens, and many receptors for vasopressin are present in the ventral pallidum. The activation of these two receptors by oxytocin or vasopressin during sex increases the release of dopamine, which has a central role in regulating the reward pathway. Dopamine is a molecule that triggers a "feeling good" sentiment after the rewarding experience, as well as after taking drugs such as cocaine or amphetamines. Just as drugs make us "high" because dopamine is released in our brains, dopamine makes both male and female voles "high" during sex.

At least in voles, it seems that dopamine does not only make a sexual encounter a pleasurable experience but also puts this enjoyable experience into the context of the partner. With drug users, the context of where and how they took drugs is very important, especially in relapse. Seeing the equipment for injecting yourself with heroin triggers an intense craving for the drug in drug addicts, as this triggers an emotional memory and desire. Dopamine released in the prairie vole brain during mating triggers a similar response. After experiencing being "high" from dopamine released during the sex, voles want to stay with their partner. They remember the gratifying experience of being with this particular vole, not just with any vole, even though different voles look indistinguishable to us.

Being "high" is a pleasurable experience for voles and humans. Voles consequently want to spend their time with their first sexual partners, with whom they experienced the euphoric effects of the dopamine during the sexual encounter. So, for voles, it seems to be all in pleasure and addiction why remain with the same partner. Similarly to blocking oxytocin and vasopressin, blocking dopamine's action during a sexual encounter in prairie voles prevents the attachment in both males and females, thus confirming dopamine's role in this process.

One brilliant scientific experiment has more or less confirmed the role of vasopressin receptors and dopamine in the male vole faithfulness. Montane and meadow vole males do not stay with their partners after mating as they have very few vasopressin receptors in the ventral pallidum. However, scientists artificially introduced vasopressin receptors into the montane voles' brain cells in the ventral pallidum. Consequently, polygamous montane vole males became suddenly monogamous and lingered with their sexual partners after sex. This experiment is the most definite proof that simple interaction between two hormones or neurotransmitters, vasopressin and dopamine, and their receptors in specific locations inside the brain, regulates monogamy in voles.

Whether the same is true in other species is not yet known for sure. We have some proofs that vasopressin in the ventral pallidum might have a similar role in some human relatives. Small monkeys, marmosets, are monogamous animals. Males and females stay together, and both moms and dads take good care of their children. Like prairie voles, male marmosets have an ample amount of vasopressin receptors in their ventral pallidum. In contrast, male rhesus monkeys, promiscuous monkeys, have very few vasopressin receptors in the ventral pallidum, just like montane vole males.

Nobody has yet looked in men's brain, but ventral pallidum is activated and brain cells release vasopressin during sexual arousal in men. We do not know whether this has anything to do with either monogamy or love, but it does suggest that vasopressin and ventral pallidum likely have some roles during sexual encounters also in human males. Whether vasopressin and ventral pallidum are essential in humans for sex, attachment, or, eventually, love is unknown. But like in monogamous animals, they probably do have some roles in the feelings of affection and desire to stay with our partners. After all, it is clear that sexual intimacy is essential in sustaining long-term relationships.

But even prairie voles are not the perfect romantic couples. Although they stay together for life and grieve their partners' loss, they also occasionally engage in extramarital affairs. Like in birds, all offspring in the same litter sometimes do not belong to the same father even in faithful prairie voles. There are differences between individuals among male prairie voles, just as in humans. Some males are great loyal partners, staying close to their spouses all the time. Other males remain with a female but would not pass an occasion for some fun with another gal if their partners are away on some business.

Vasopressin receptors are very similar in the prairie and montane voles and in all prairie vole males, regardless of their faithfulness. Therefore, it is not how the receptors are built, what determines where they are active, and how they cause differences in behavior. Genetic analyses showed that genes for the vasopressin receptor differ between voles in an area that regulates the gene's activity. This difference is likely responsible for the differences where the gene is active, and consequently, for the presence or absence of bonding between males and females. Prairie voles have an additional long stretch of the DNA in the regulatory region of the gene for the vasopressin receptor. This extended region is absent in the montane and meadow voles, and it is this region that is responsible for the presence of vasopressin receptors in the ventral pallidum in the prairie, but not in montane voles.

Interestingly, there is a difference in this genetic sequence between prairie vole males. Males that are truly faithful to their partners have a longer stretch of this DNA in the vasopressin gene. In comparison, males who are more prone to occasional adultery have a shorter stretch of the DNA. In voles, it is, therefore, this regulatory region of the DNA that determines who is a faithful partner and who is Casanova among the prairie voles.

Vasopressin receptor is similar between voles and humans, but the regulatory region of the gene is not. There are at least three different variants of the vasopressin receptor gene's regulatory region in humans. Although such experiments are still in their infancy, some reports showed differences in the vasopressin receptor gene's regulatory region that might correlate with our behavior. One study reported that a particular variant of the vasopressin receptor gene is associated with more marital problems. Men with such alterations of the gene were more often single or experienced a divorce. In another study, though, scientists did not find a relation between differences in the vasopressin receptor gene and adulterous behavior in men but, surprisingly, found such correlation in women. On the other hand, oxytocin and its receptor were not found to correlate with marital problems or extramarital affairs. So, at least for now, we cannot blame the genes for adultery, and we cannot perform genetic tests to predict who will remain a faithful partner and who will likely seek new intimate adventures.

All these studies are in their infancy, and there are several essential points to remember. Firstly, and very importantly, not only for understanding love but also general science, correlation is not causation. If scientists found the relationship between vasopressin receptor gene structure and adulterous behavior, it does not necessarily mean that genetic differences cause such behavior. We have numerous funny examples of why we should not equal correlation and causation. There is a solid correlation between the number of people drowning in the pool per year and the number of new movies with Nicolas Cage as an actor in the same year. A robust correlation exists between the consumption of cheese in the USA and the number of people dying tangled in their bedsheets. Similarly, the number of people dying after falling from the wheelchair is strongly correlated with a US consumption of the sour cream.

We all know that watching Nicolas Cage in a movie does not cause drowning in the pool, regardless of how bad the film might be, and eating sour cream does not make you fall out of the wheelchair and die. So equally, we cannot conclude from two studies that the vasopressin receptor gene has a definite role in regulating our marital behavior.

Most importantly, we are conscious beings, and our frontal cortex has control over our actions. So, we and our conscious brain are responsible for our behavior, and we cannot blame it on the genes. Yes, similarities between animals and humans suggest that oxytocin and vasopressin likely play some roles in the regulation of our social behavior. They might have a limited role in why some of us are firmly attached to our partners and others are not. But so far, there is no genetic test, and it is unlikely there will ever be, that will tell whether somebody will be a loving, faithful partner for many years to come or whether our partner might change his or her mind and seek greener grass in other pastures.

So, is faithfulness written in our genes and hormones, like it seems to be in voles? Undoubtedly, it is not only vasopressin and oxytocin that are partaking in the attachment and love. There are many other genes involved. Even in voles with much simpler brains, some genes are important before vasopressin and oxytocin can act, and genes that work after vasopressin and oxytocin perform their duty. Before the vole couple meets for the first time, genes must ensure sexual maturity and fertility. Some genes are responsible for voles' sexual receptivity, genes that provide recognition between the partners, and many others. Once vasopressin and oxytocin trigger the feelings of attachment, other genes spark cuddling behavior and parental instinct.

Clearly, it is not only vasopressin and oxytocin that are involved in the feelings of attachment, but, undoubtedly, these two hormones have crucial roles, at least in voles. Whether they are also involved in human love, and to what extent, we do not know. But what we can say for sure is that neither vasopressin nor oxytocin are a magic love potion that will make somebody fall in love with us. Vasopressin and oxytocin might trigger feelings of attraction and attachment, but building a strong and lasting relationship requires much more. It requires trust, confidence, common goals, social cohesion, and much more. Although vasopressin and oxytocin might facilitate some of these feelings, our frontal cortex, a conscious brain, will ultimately determine whether we will stay with somebody for life or not.

# Chapter 20
# Are There Romeo and Juliette Among Chimpanzees?

*One thing I had learned from watching chimpanzees with their infants is that having a child should be fun.*

Jane Goodall

Although many would agree that we are somewhat similar to apes, most of us forget that we are apes. Biologically, we are members of the group of apes, tailless monkeys. Chimpanzees, bonobos, gorillas, orangutans, and gibbons are our closest natural and evolutionary relatives. We are different because we developed along different evolutionary pathways, with chimpanzees and bonobos for more than 5 million



G. Majdic, *Soul Mate Biology*, Fascinating Life Sciences,
https://doi.org/10.1007/978-3-030-67212-6_20

years. More than 6 million years ago, we split from a common ancestor with gorillas. Around 10 million years ago, orangutans and humans started our evolution along different paths. The earliest split within the apes occurred about 18 million years ago when gibbons started their evolutionary trajectory.

In contrast to the common belief in popular culture, we did not develop from the chimpanzees. Chimpanzees and bonobos are our closest biological relatives. We all evolved from the same ancestors, but we are also evolutionary equally old species. Evolution is not a linear process, one species developing from another. All living organisms are constantly changing, and therefore diverging during evolution. As much as we have changed during the last several millions of years, so have the chimpanzees and bonobos. They are as much different from our common ancestor as we are.

To say that we have developed from a chimpanzee is like saying we have developed from our cousins. Our cousins and we have the same granddads and grandmas, great grandfathers and great grandmothers, and ancestors further back. Still, we are contemporary relatives and have not developed from each other. Chimpanzees, gorillas, and orangutans have the same right to say they have developed from humans, just as we often say that we evolved from the chimpanzees. We can only guess whether chimpanzees would be appalled by such a claim as much as some members of our species are horrorstruck by the idea that we have the same ancestors as other apes. Even though orangutans, chimpanzees and gorillas might have much better reasons to be unhappy with us as their relatives, considering how we are destroying their habitat and endangering their survival.

As we split apart so long ago, this also explains why we are so different in some respects, including sex and family life. Five million years is, in evolutionary terms, a period long enough to create two very distinct species. These differences are not only in the appearance but also in the reproduction and social behavior, which is certainly very different between chimpanzees and us. Sometimes, we try to simplify things and pretend we can learn much about our behavior from studying other apes. Such ideas are, to a large extent, a futile effort. We are our own species. We cannot find excuses for our behavior in the often violent nature of chimpanzees or the promiscuous sexual behavior of bonobos. We alone are responsible for our behavior.

None of our closest relatives, chimpanzees, bonobos, gorillas, or orangutans, are monogamous. Each species has developed its own habits, social life, courtships, and mating practices, very different from our social patterns and love. They do not cherish love, and they do not live in unisons of one male and one female. It seems that even though the apes are our closest relatives, we cannot learn much about love from them.

This difference does not mean that chimpanzees, bonobos, gorillas, and orangutans do not have emotions and do not develop close relationships. Mothers take great care of their infants, akin to love between human mothers and their children. There is undoubtedly something like a friendship present between the members of the same troops. Some individuals get along better with some peers than others. Romantic love, however, does not seem to exist in our closest biological relatives.

But, in contrast to chimpanzees, bonobos, gorillas, and orangutans, monogamous are our most distant relatives among the apes, gibbons.

Gibbons, apes living in several Asian countries, are in many characteristics more similar to other species of monkeys than to apes. They spend most of their time in the trees, hanging from twigs or sitting on thick branches, and sing in loud, harmonious duets. Although they can walk on the ground with two legs, they usually graciously swing from one branch to another, using their powerful and long arms. Their arms are 1.5 times longer than their legs, and they can propel them to the leaps over 15 m or 50 feet long. With these jumps, Gibbons can move through the thick Asian jungles with an admirable speed of over 50 km/h (35 mph). Gibbons have one characteristic that categorizes them among us, apes, and not monkeys. Unlike monkeys that can hang from the trees with their tails coiled around the branches, gibbons do not have tails.

In addition to being tailless, they share one more thing with us—gibbons are monogamous most of the time. There are 18 species of gibbons (some are very rare and threatened to be extinct very soon), and all of them spend most of their lives in seemingly harmonious monogamous relationships between one male and one female. When we observe adult gibbons, we almost exclusively see them in pairs, composed of one bloke and one gal. For years, gibbons were the symbol of faithful love. Scientists, who observed them for short periods of a few months, always saw them in pairs. These earlier observers consequently classified them, partially based on the experiences or desires with our species, as strictly monogamous animals. Scientists presumed that young gibbons, once they are kicked out of the house by their parents, find partners, establish a new core family, and live happily ever after.

More extended, more thorough studies revealed that gibbons' social and love life is more similar to real human behavior, not the one we idolize in books, poems, and movies. Both male and female gibbons are not impervious to an occasional extramarital affair. Gibbons stay together with one mate and most often have sex with this partner. But, if the occasion arises, they do not spend too much time thinking about whether to engage in an adulterous affair.

Gibbons usually don't commit for life. Although they are almost always in pairs, these pairs change during their lifetime. If there is a widow or widower in a neighboring territory, gibbons might decide to try a new relationship. They would leave their old partner and begin a new romantic liaison with a novel partner. Perhaps the grass growing in a neighbor's garden is greener. But divorced partner usually does not stay alone to lament its destiny for a long time. Most often, a lonely gal or bloke soon finds a new single partner without using Tinder or advertising his status on Facebook. Happily ever after starts again, with the exception that ever after often lasts only a few years.

So, gibbons' love life is very much like ours. We do not know whether they cherish romantic love as we do. What we do know is that they spend most of their time in monogamous relationships. Still, these relationships break, and new relationships are formed, just as in our society. Gibbons cheat on their partners if a situation arises. Unlike swans, who most often divorce if the couple is childless, fertility does not seem to be a factor in the gibbon's divorce. The natural explanation for such behavior might be merely increasing the combination of genes in the next generations by having children with different partners.

Gibbons are probably not staying together because of love but more likely due to opportunistic reasons. Two gibbons can better protect their territory and raise their children. They are very territorial animals. A couple can control the large household and prevent intruders from entering their estates much easier and more effectively. A gibbon pair usually manages between 50 and 100 acres of real estate, and both males and females vigorously protect their land. They move around the jungle, checking for intruders. They claim ownership over their territory by singing together. Male and female gibbons often sit close together on a thick branch and sing in a harmonious duet. With singing, a couple tells other gibbons that the territory is occupied and they should find themselves a home somewhere else.

Males sing a different song than females, but both songs together join into a melodious tune. When with the family, younglings often join in singing, but immature gibbons always sing a female vocal, regardless of their sex. Why young males sing a female song instead of learning how to sing like a male is still a mystery. The most likely explanation is that if they would sing a male song, their father might recognize them as a usurper and threat and may attack them. So, young blokes begin to sing a male song only when they are detached from their family and become independent.

Quite the opposite from gibbons are orangutans, orange furry forest-humans (word orangutan means forest-human in the Malay language). Orangutans live in a thick jungle in Sumatra and Borneo islands in Southeast Asia. Adult male orangutans are most of the time hermits, living alone, avoiding contact with other members of their species. Male patriarchs control their territory and strictly forbid other males to mess with the females living within their range, even though they are not socializing with the females. Females and young orangutans are more social. Orangutan mothers are second only to humans when it comes to love and care for the young. Children stay very close with their mothers for up to 10 years and are often not wholly independent until they are 15 or 16 years old. So, although adult males are solitary loners, adult females and adolescents of both sexes are socializing, not only mothers and children but occasionally also unrelated adolescents and young males.

When it comes to mating, however, the real boss is the lonely hermit. Male orangutans come in two forms. Adolescents and many adult males look similar to females. They are smaller than fully mature males and without any special appearances. Adult, mature males are much larger. They are called flanged males. They have very prominent cheek pads on the sides of their faces and sizeable pendulous throat sacks under their chins. These males often produce loud sounds that can be heard up to a mile away. With these sounds, they attract females and warn other males to stay away.

The bodily characteristics of a flanged male are not connected with age nor with sexual maturity. Many adult males are sexually mature and fully capable of reproducing, but they do not have the appearance of a flanged male. Although it seems likely that all males will eventually reach the flanged stage, the age at which they do is very varied. It is not known what regulates the maturation of males into the flanged males. One hypothesis suggests that the singing of a flanged male somehow, through hormonal regulation, prevents other males in his range from entering the final process of maturation.

Flanged males are the real bosses when it comes to sex. They are the blokes within a territory that mate with most females. The flanged male is so popular that females are seeking him and not vice versa. Loud singing by flanged males attracts females in heat. When given a choice, females will always choose a flanged male over a non-flanged male.

Flanged male can just sit in his space and wait for the ladies to approach him. It is usually the girls that perform most of the courtship behavior. If the chief is pleased with a girl, a couple may stay together for several days. The female stays close to the male, and they offer each other food. It is also the female who usually initiates sexual contact by rubbing and sometimes orally arousing the male. Even when it comes to sex, it is usually the female who has to do most of the work. Flanged male sometimes just lays on his back and waits for a female to climb on a top and do all the work to make him ejaculate. Often, he looks away during sexual intercourse, like he is not interested in what is going on and just waits for the gal to finish whatever she wants from him.

Younger, unflanged males, however, are sexually mature and often horny. Yet, they cannot impress the females with their inflated cheeks or loud sounds. They are not attractive to the girls. The only way to get sex is to become violent and force sex with females in heat. Indeed, in orangutans, something akin to rape is a fairly common occurrence. If an unflanged male during his wanderings in the jungle encounters a female in heat, he will often force her into sex to get his few minutes of pleasure. But he must be careful not to be close to the flanged male. The boss does not gladly share his harem. If a flanged male sees an unflanged male raping one of his females, repercussions are severe in the form of a merciless beating.

Although orangutans' maternal love might be as strong as humans', there is hardly anything akin to love between adult furry orange apes. Orangutans, especially young, are very cute and adorable to us and probably to orangutan mothers. But adults between themselves are not the most caring individuals. Adult males often carry signs of previous fights such as scars, missing fingers, or even lost eyes.

Gorillas are much more social than orangutans. They live in troops, usually composed of around ten individuals, although troops could be smaller or, rarely, more numerous. Several males and females are members of the troops, but, like in orangutans, a single male monopolizes the troop when it comes to sex. Also, similar to orangutans, this prime male is different from the other males. He is the largest male in a troop and has distinct gray hair on his back. Hence, he is called (by us) the silverback.

Silverbacks are taking care of their troop, protecting members, and leading them around. In return, they get the best food. Even when the food is scarce, silverbacks receive the royal share. Not only the food, they also claim their royal rights over the female members of the troop. Silverback is the only male in a troop allowed to mate, and if this rule is broken, the consequences are dire for the usurper. But the temptation of having sex and transferring one's genes into the next generation is alluring. Younger males in a troop occasionally sneak out of sight of the silverback and quickly mate with the willing female. Infants in the troop, therefore, do not belong all to the silverback, increasing the genetic variability between the troop members.

Within the troop, we can observe the friendship between the members, but there is little evidence for any emotional attachment between the silverback and his harem. Females initiate the contacts with the silverback and seduce him into mating when they are in heat. He willingly has sex with the seducing female, sometimes in a missionary position, male and female facing each other, but he does not show much affection.

Within the females, there is a definite hierarchy. High-ranking females, usually the oldest troop members, are better protected by the silverback and get better and more food. Similarly, infants of higher-ranking mothers are better protected and have a better chance of survival. Children's protection is especially crucial for gorillas, as infant mortality is very high, with only about 60% of all babies surviving.

Female gorillas show some unusual behavior to remain dear to the silverback's heart and better provide for their children. Even when they are already pregnant, females pretend to be in the heat and lure silverback into the sex without the prospect of pregnancy, which is a rarity among animals. So why are they doing it? One possible explanation is that having sex with the silverback helps high-ranking females protect their social rank. Not by seducing the silverback into monogamy but by preventing other females from entering the quotation. The highest-ranking female would sometimes offer herself for sex to silverback whenever another fertile female came into the heat. By having sex herself, despite already being pregnant, she prevents another female from being impregnated and becoming a mother. Sometimes, a subordinate female that cannot have sex with a silverback even leaves the troop. A dominant female thus gets rid of her competitor.

Although not particularly parental, silverbacks and other males spend time with the infants. Silverbacks protect the infants, play with them, and ease the tensions between older and younger siblings. Just one stern look is enough to stop the older males from bullying the younglings.

Looking through the eyes of evolution, parental care by males, especially for children who are not ours, is a waste of energy and resources. Yet, all male gorillas within a troop spend time playing with children and care for them even when they know they cannot be theirs. They might be doing it to impress females. Female gorillas seem to note which male is spending more time and takes better care of the infants. Such males are more often chosen for sneaky sex behind silverbacks' backs. They have thus a better chance of transferring their genes into the next generations.

So fatherly love in gorillas might not be a real emotional attachment to the cute babies, but an evolutionary advantage for males in obtaining females. So, they behave a bit like Will in Nick Hornby's novel *About a Boy* (and the movie with the same title), who is pretending to be a caring father, inventing a fictional son to be more successful with women.

Chimpanzees are social animals like gorillas, but their society is somewhat less autocratic in comparison to gorillas. They live in large communities, often consisting of 100 or more individuals. An alpha male leads the community, but much social life is going on separately within these large communities. Smaller groups of friends, called parties, often form and are disbanded. These groups could be made of friendly individuals, just wanting to socialize together, or comprise several younger adult males, joining together in planning a coup to overthrow the alpha male. Alpha male, however, is closely watching, and he usually has his group of buddies ready to protect his and their high positions within the community. These friends of the alpha male are often his siblings or males he grew up with during his adolescence.

When it comes to mating, chimpanzees are very promiscuous animals. They follow numerous mating patterns. Although the higher-ranking males usually have sex more often, there is a time and place for lower-ranking males. As parties form within the communities, these parties often separate from the leading group, and alpha males could not control all smaller parties. Very often, female chimpanzees have sex with multiple male partners within a short time. With such promiscuous behavior, it is essential to produce as many sperm as possible to have a better chance to be a father. Chimpanzees' testes are consequently very large, producing vast amounts of sperm.

Although the rules are more relaxed in some larger troops, alpha males try to be similarly oppressive as gorilla's silverbacks in many chimpanzee communities. In such troops, most of the mating is performed by an alpha male, who is carefully guarding his females, as well as males. He is often helped by friends, and these associates are rewarded not only by food but also by more sex. When it comes to paternity, the second highest-ranking male, usually a close friend of an alpha male, is second only to the alpha male with the number of children within the community.

Females have two ways to tell the blokes when they are ready to have sex. When in heat, their genitals become red and swollen, a pretty obvious visual sign. If this is not enough, ladies are also inviting blokes to approach them with a specific calling. Red and swollen genitals are a sign for every bloke in a gang, but the girls can show their preferences with calling. Usually, they prefer higher-ranking males, and they more often call toward them than to the lower-ranking males.

Chimpanzees are very promiscuous, and males could never know who the father is. Sexual interactions occur between different troop members, but having sex with higher-ranking males raises your position in the community. With the higher position comes better protection, better access to food, and better protection for your babies. So not entirely unlike in our species, chimpanzee females know how to use their ladies' charms to advance their position within society. Girls have worked it out quite well, and sometimes, they even offer sex to horny blokes in exchange for the succulent piece of meat. Perhaps the oldest profession is indeed the oldest and by a much larger margin than we usually think.

But it is not only the females who are choosy. The blokes, including the alpha male and his gang, have their preferences as well. Quite different from many human societies, chimpanzees do not cherish inexperience when it comes to sex. Females who have been mothers before are more attractive to males. Higher-ranking males, if given a choice, will always choose older, experienced females. Probably not due to their experience with sex. Being a mother previously is a pretty sure sign of fertility. Mating with such females thus provides a better chance for being successful at transferring ones' genes.

With higher-ranking males trying to control the females within the community and girls preferring higher-ranking blokes, low-ranking males might have trouble getting laid and transferring their genes into the next generations. Such low-ranking blokes sometimes run out of options, and they become aggressive. They threaten the females or lead them away from the community to have sex with them, although the sex is never as violent as in orangutans, as females often do not resist too much.

In general, it seems that there is no sign of anything like love in chimpanzee communities. But occasionally, a bloke and a gal separate from other chimpanzees and seek solitude and togetherness. They venture away from other chimpanzees and spend a few days together, having sex often and only with each other. But as romantic as this might sound, it is unlikely it has anything to do with the amorousness. A more likely explanation is that the male simply wants to ensure he is the baby's father. By leading the female away, sometimes forcefully, the male prevents the female from having sex with the other blokes. So, in reality, it is more like a sexual kidnapping than romantic love.

Indeed, rather than in love, we might find more similarities between our and chimpanzees' behavior in violence. Chimpanzees can be very ferocious. Blokes within the community often fight for food or girls or to achieve a higher social rank. A male falling out of favor with high-ranking males within the community is sometimes savagely mauled. Even more forceful are males toward the intruders into the community's territory. While female chimpanzees sometimes move between communities, males do not tolerate any male intruders. They viciously attack any male wandering purposely or accidentally into the area of the foreign community.

A split in the chimpanzee community led to a 4-year war between two chimpanzees' communities in Tanzania in the 1970s. After the death of an alpha male, a weak successor took over the throne, but two strong blokes challenged his rule. Each party had its supporters, and 4 years of fighting broke out, ultimately ending with the death of all males in the defeated company. As far as we know, the chimps are the only other animals besides humans known to wage long-lasting wars.

The last group of apes is the bonobo, close relatives of chimpanzees, living in a small area in Central Africa. Bonobos look quite similar to chimpanzees, but their behavior is very different. They are sometimes referred to as hippy apes due to lack of violence, especially in comparison to chimpanzees, and widespread habit of having sex often within their communities. While chimpanzees solve most conflicts by force, bonobos use a different approach, and sex is very prominent in their relationships. Whenever tensions arise between males and females, or between two females or two males, sex and not brawl seems to be the right answer in the bonobo community.

Sex is present in many different forms within the bonobo society. Males and females engage in heterosexual sex, including oral sex and face to face position.

Genital rubbing is very common between two females. Two blokes often stand with their backsides together, rubbing their testes, or perform penis fencing by facing each other and rubbing their erect penises. Whenever tensions arise within the bonobo community, bonobos do not become violent but relieve tensions through sex. Before feeding, they would engage in sex, which seems to ease the tensions who will get more food. If two females clash, instead of going on and on about who is right, some genital rubbing solves all the open questions. If two males compete for the same female, they might end with penis fencing instead of having sex with the female. And like chimpanzees, females offer sex for food.

Another peculiarity of bonobo society is matriarchate. In chimpanzees, the strongest and smartest male is the leader of the community. Bonobos instead rely on female leadership, and perhaps this explains the peaceful nature of their communities. If male tries to assume too much power, he is quickly put in ranks with dominant females, true leaders of the bonobo society.

Although sex is so vital in bonobos, it is separated from love and attachment. It probably serves primarily as a means to ease the tensions, to hold the community together, but it has little to do with the attraction between two individuals.

Although we often think we can learn a lot about ourselves by studying apes, this seems unlikely when it comes to love. The only apes that live in monogamous relationships are our most distant relatives, gibbons. Other apes show patterns of social and sexual behavior that might resemble our nature in some respects, but not the one we cherish and value. Apes are often violent as we are; chimps even wage wars. But when it comes to love, it seems that our species made a long step forward during the last 5 or 10 million years of history, although we do not know why and how that happened.

# Chapter 21
# Emphatic Rats

*Ah! could I behold the face of my beloved once more, I would give thanks unto my God until the day of resurrection.*

Inscription on the Tomb in the Pakistani City of Lahore

No one knows who is buried in the grave with this inscription in Lahore in Pakistan. An Indian legend says that this is the tomb of Anarkali, a court dancer girl who was in love with Prince Saleem, the son of the mighty mogul Sultan Akbar. According to the legend, Saleem was so enchanted by Anarkali's beauty that he went to war with his own father when the father denied him the right to marry an ordinary girl like Anarkali was. Eventually, his father's army defeated Saleem, and he was

 
G. Majdic, *Soul Mate Biology*, Fascinating Life Sciences,
https://doi.org/10.1007/978-3-030-67212-6_21

condemned to death unless he renounces his love for Anarkali. Prince Saleem refused to do so, but Anarkali sacrificed herself for the sake of her lover, gave herself into the sultan Akbar's hands, and was entombed alive. This story is probably just a story without much historical basis. But it does show that romantic love and thoughts of everlasting love existed in human societies for a very long time and in different parts of the world.

Some claim that love is a recent invention of western societies, but this is unlikely. Romantic love exists in most, if not all, human cultures. If ideas of romantic love would not live in the past, would we have a story about Snow White meeting a prince and Sleeping Beauty being awoken by a kiss of true love? Would Orpheus venture into the underworld to bring back his Eurydice? Would Shakespeare write his love sonnets? Would Tristan die of despair, waiting for his beloved Isolde?

Love is undoubtedly accompanying us from the dawn of civilization and not just in western culture. Famous Persian poem Majnun Layla talks about immortal love between Layla and Qays. Shah Jahan built a beautiful monument, Taj Mahal, in the remembrance of his beloved wife, Mumtaz Mahal, in Agra in India. Prince Saleem from Lahore went to war against his father for the right to be with his beloved Anarkali. In Chinese mythology, a famous love story talks about Zhinu, the weaver girl, and Niulang, the cow herder. The lovers were not allowed to be together and were banished to separate sides of the river. For Chinese, they are presented on the night sky by two bright stars, Vega and Altair, separated by the Milky Way. And this story is at least 2600 years old!

So, love seems to be embedded in our societies and our souls from the dawn of civilization. Yet, our closest relatives, chimpanzees, bonobos, and gorillas, do not show anything resembling romantic love. So how did love evolved and emerged in humans?

Could love arise from love in animals? Love is an emotion. As we cannot measure or monitor animals' emotions, it is impossible to understand whether animals experience something akin to love. Yes, animals can show affection, friendship, and bonding. Prairie voles show something similar to grief after the loss of the partner. Gibbons sit together in a tree in what looks like a harmonious relationship. But love is more than that. Love is not a behavioral pattern. Love is a feeling. A desire to be with our loved one. A dream about our loved one. A sense of anxiety when we are separated from our loved one. Love is a cognitive and emotional relationship with another human being. As such, it also requires the awareness of self, empathy, foresight, and cognitive abilities. Hence, we cannot talk about love in animals, and we cannot envision that love in humans has simply evolved from the passion in animals.

Empathy was considered to be unique to humans. However, we know now that empathy is also present in animals. Elephants mourn their dead; dogs seemingly show signs of compassion toward humans in distress. Numerous studies have convincingly demonstrated empathy in rats and mice. Rats push the lever in their cage to help a drowning rat nearby. When given a choice between a tasty snack and saving another rat from drowning, rats will give up on a snack and save their mate from drowning by pressing the correct lever. Mice show signs of pain when they observe other mice being in pain. Interestingly, mice only show an emphatic response when a mouse they are acquainted with is experiencing pain, but not when they see an unfamiliar mouse being hurt. This lack of sensitivity might not mean that mice do not feel empathy toward unfamiliar mice. Rather, the presence of a stranger, an alien mouse, might trigger such intense stress that this prevents the signs of empathy. Indeed, if stress hormones are blocked in the observing mouse, such a mouse shows empathy when the alien mouse is suffering.

Rats learn quickly to press the lever to release another rat trapped in the water or tiny cage. Unlike mice, they will do it also for the rat they were not previously acquainted with. But they will help a fellow rat only if this rat belongs to the same breed, or strain, as it is called in rats.[1] White Wistar rats will free a fellow trapped Wistar rat even when the alternative to releasing a rat is a tasty chocolate chip. However, if they have to decide between freeing a rat that looks different or getting a chocolate chip, rats will rather go for a snack.

But things change if rats have a chance to become acquainted. If two rats from different strains (with a different coat color) live together for 2 weeks, each one will help another in troubles, even though they have different coat colors as they belong to different strains. But this gets even more interesting. If pups from one strain, let's say with a brown color, are fostered by a white mum, such fostered brown rats will empathically release white rats as adults, even when they are not familiar with them. These suggest that even in animals, as in humans, empathy has an important social context. Just as we are more likely to help a member of our social group, so are rats.

Cross-fostering experiments in rats suggest that difference in empathy toward a familiar person or a stranger is not embedded in our genes. Instead, it seems to be a result of our upbringing and early socialization. This has important implications for our societies. It means that at least partially, negative emotions and lower empathy toward strangers and people that look different from us are not inherent in our genes. They are results of our upbringing, and we can influence such emotions in our children. So, we have the tools and powers to make our societies better and more inclusive through our children's upbringing. If we teach our kids to be emphatic and inclusive, perhaps future generations will be more peaceful and will build a better, kinder world.

All these studies about empathy in animals indicate that empathy is not unique to us, humans. It is present in animals, and we simply inherited it from our ancestors. But could love evolved from empathy?

Empathy is an essential part of love. Notwithstanding, it is much more likely we will help a member of our family than a stranger. But empathy is not love. Empathy is a neutral perception. It can be sympathetic, but it might also produce a hostile reaction. We can empathize danger and attraction, fear and aggression, hatred and affection. So, empathy is not love. It might be part of love, but empathy is a group feeling. In contrast, romantic love exists between two individuals. Therefore, empathy is unlikely to give rise to romantic love in humans.

---

[1] Laboratory rats are classified into different strains, similar to human races. They all belong to the same species, but they differ in specific characteristics like the coat color. The most well known are probably Wistar and Sprague-Dawley rats.

We often associate love with sexual desire. We even call sexual intercourse "making love." Love does trigger the thoughts of sexual relations, but love and sex are also separated. Sexual desire is present in all animals, and it is certainly not connected to emotional attachment. In animals, mating is rarely a romantic experience. There is hardly love in a praying mantis biting off the head of the male. In many mammals, sexual intercourse is a quick, often quite aggressive act.

Sex is often dissociated from love also in humans, although, interestingly, this is different between men and women. Men would often and leisurely separate sex from love, while for women, an intimate emotional relationship is often, albeit not always, a prerequisite for having sex. This difference has probably evolutionary origins in the behavior of many mammals. Most mammal males are not faithful to one female partner. Males try to mate with as many females as possible to increase the chances their genes will survive into the next generations. They do not have to take care of the brood. A better strategy is to sire as much progeny as possible and leave it to the chance that some of the genes will survive in the next generations. For females, the stakes are much higher. Females can have only a limited number of offspring, and their investment in the survival of their genes is much higher.

For a male, quick sex is enough to produce offspring. Women have to invest considerable resources during several weeks or months of pregnancy and, later, when caring for children. Therefore, females have to be choosier when searching for a partner, the father of her babies. They must choose the best available prince. The one that will provide the best genes for offspring and perhaps help them raise the children. This will increase the chance of progeny survival. Consequently, sex and love are much closely related in women than in men. When asked if you agree with the statement "Sex without love is fine," the chances that you will answer affirmatively are much higher if you are a man than a woman. Indeed, the separation between sex and love is one of the greatest and most consistent sex differences in human societies regarding sexuality.

Sex and reproduction are separated from emotional attachment and bonding, not just in humans but also in animals. Although sex in prairie voles prompts the attachment between males and females, it is just a spark that triggers the attachment and not the fuel for burning the fire.

In humans, sex is most often an integral part of love and a long-lasting, harmonious relationship. Still, love is much more than just sex. It is a desire to be together, do things together, share experiences in unison, and laugh and cry together. Our societies cherish long-lasting relationships. The vast majority of people stay in monogamous relationships for at least parts of their lives. Sex is an important but not exclusive part of such long-lasting relationships. Often, sex in committed, harmonious relationships is less important than in short, passionate affairs. Stout sexual desire can be even detrimental to the long-lasting loveable association between the two partners, as it could lead to adultery.

Sexual reproduction has its roots far away in the past and is present in the vast majority of animals. Although behavioral patterns implicating sexual desire might be more recent evolutionary inventions, they are present in most mammals. On the other hand, bonding and attachment, yet alone love, are not, which makes sexual desire unlikely predecessor for the evolution of love in our species. It is therefore very doubtful that sex drive was an evolutionary origin of love.

Therefore, we have one last option. There is one relationship that includes empathy, emotions, and attachment. There is no such love as motherly love. As Agatha Christie said in one of her books: "A mother's love for her child is like nothing else in the world. It knows no law, no pity. It dares all things and crushes down remorselessly all that stands in its path." Mother and child bonds are present in all mammals and many other animals. In mammals, this special bond between a mother and an infant child is needed for the survival of the offspring, as newborns cannot survive alone. From birth, mothers develop a close relationship with their progeny. Human and animal babies are instinctively attached to their caregivers soon after birth. Similar to the attachment between the adult male and female voles, oxytocin and dopamine have a central role in the bonding between the mother and her brood. This mechanism is conserved between animals as in every mammalian species examined; oxytocin released during the labor and lactation has a central role in developing the attachment between mother and her offspring.

In humans, attachment between parents and children is often explained by four domains:

- Proximity seeking (desire to be close to parents).
- Separation distress (uncomfortable feeling when being away from parents).
- Safe haven (parents provide comfort).
- Secure base (exploring a new environment within a safe distance from the parents).

A fulfilling relationship between the two grown-ups is very similar. We desire to be close to our partner and feel lonely when we are separated. We seek comfort in an embrace of our partner when we are sad, and we try to keep close even when we are separated.

Does this mean that mother and infant relation is the same as love in humans? Certainly not. Mother and child attachment is present in all mammals and in many other animals, but hardly anybody would consider caring for brood in ants or beetles as love. Indeed, we cannot understand or measure the feelings and emotions of animals. If we look at apes, mothers undoubtedly develop a special relationship with their babies. This relation involves mutual understanding, empathy, care, and protection. Still, it is difficult to call this love.

Romantic love in humans has likely origins in the bond between mother and infant, but it has also progressed beyond it. Maternal care has one aim: ensuring the survival of the offspring and, thus, one's own genes. Romantic love is more complicated. Biologically, it is likely that the original reason for love came from the same premise—better protection of the children and enhanced chances of gene transfer into the next generation. But love is more than that. It is sharing the experiences, enduring hardships, and enjoying beautiful moments in unison of two partners. Indeed, being childless is often a reason for divorce. Still, many couples persist in a loving relationship without ever caring for children. So, love in our species is more than just care for the kids. Love requires higher cognitive abilities and has thus likely developed alongside the enlargement of our brain.

The most crucial difference between bonobos, chimpanzees, gorillas, and us is a much larger human brain. A larger brain is responsible for higher cognitive capacity, communication by language, analysis and control of the environment, foresight, and planning future actions. Admittedly, some of these are present in apes and, to some limited extent, in some other animals. Still, they have not developed as much as in our species. However, these abilities came with a price. Our brain is very large, and as a consequence, human babies are born with a relatively undeveloped brain. More mature babies with larger brains and heads could not be born, as they would not pass through the birth canal. Consequently, human babies are born more helpless than many animal babies, and the human brain develops massively after birth.

Therefore, human babies need more, better, and more extended protection. Such protection can be much more manageable by a couple than by a single mom. Two parents can take care of each other, and one member can provide the food for both. They can share chores and protect the baby in a safe haven while the other parent searches for the food. To ensure all this and subsequently increase the chances for babies' survival, evolution invented a unique form of attachment and bonding, a feeling we call love. Love is a relation between two individuals with the primary aim to care for the children. But it is not the intuitive protection of a helpless infant. Nor is it a physical relationship but a neural and cognitive connection.

Love involves understanding others' thoughts and desires, longings, and aspirations. As such, it demands self-consciousness, self-awareness, and the ability to project our thoughts and perceptions into another person. The ultimate biological reason for why love exists seems to be the protection of the human child.

The survival of the species began to depend on both parents caring for their young several million years ago. Alongside the occurrence of a large human brain, this initial instinct to protect the child has developed into conscious behavioral patterns. These include commitment, mutual care, empathy, and many other feelings, as well as sexual desire, attraction, and feelings of attachment. Love and commitment, aimed to protect the children and thus one's genes, have become embedded into our genomes and developed into romantic or romanticized love. An attachment between the mother and infant has evolved into romantic love, as this provided better chances for the survival of the species.

Evolutionarily, love might have also evolved as an instrument to ensure long-lasting commitment. If only rationality were imperative in our lives, we would leave our partner when somebody who is better looking, younger, stronger, or richer will come around. We will abandon our partner when he or she becomes sick or poor. Admittedly, this does happen, but not as often as one would predict if rationality would prevail. Love is dwindling the rationality. It ensures that we stay together in a committed relationship also in times of troubles.

Love is an emotion that forces us into staying together, even when somebody more desirable comes along, as we are committed to our partner through love. This commitment is essential for several reasons, but biologically and naturally, the most important is the upbringing of children. In the past, during the evolution of our species, children had a much better chance of survival if both parents cared for them. Therefore, parents who showed a higher level of commitment to one another produced viable progeny more often. Their genes passed into the next generations easier. Gradually, this has, over millions of years, evolved into what we call love today.

# Chapter 22
# Brain in Love

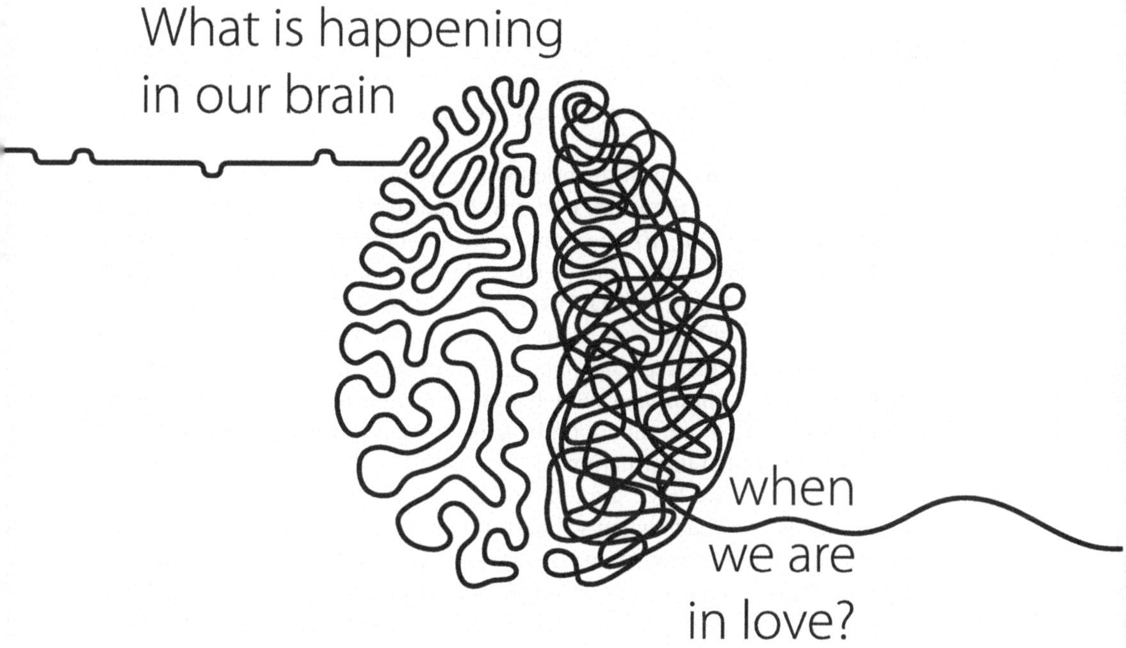


*The brain is the most outstanding organ. It works twenty-four hours a day, seven days a week, 365 days a year from birth until you fall in love.*

Sophie Monroe: Afflicted

Every one of us has a big, gray, slithery blob of tissue inside our heads. Although this tissue looks quite unappealing, it is the most crucial part of our bodies. Our bodies would not function without it. To paraphrase the famous philosopher Rene Descartes who said: "I think therefore I am," we can say, "My brain thinks, and therefore I am." The brain is the central processing unit of our bodies. Computers have microprocessors to do everything we want computers to do. Similarly, billions of nerve cells, forming a brain, control everything we do. From unconscious



G. Majdic, *Soul Mate Biology*, Fascinating Life Sciences,
https://doi.org/10.1007/978-3-030-67212-6_22

decisions, when are we hungry and when are we sleepy, to higher cognitive functions, reasoning, foresight, and others. Everything we do and everything we think comes from the brain and so do the feelings of love.

The brain is our most complicated organ. We know pretty well how the heart beats and transports blood around our body. We understand how the gut digests the food and sends the nutrients into our body. We learned how kidneys produce urine and remove harmful substances from our body. However, the brain is often considered the last frontier in biology and medicine, as we still understand very little of what is going on in our brains. We know how brain cells use electricity to talk to each other. We have learned for many parts of the brain what they do. We understand how the brain triggers the movement of our muscles. We partially understand how memory works.

But so far, we do not know how the brain works as an entity. We are slowly learning the function of each of the brick in the wall, but we lack an understanding of how the wall stands upright. If we compare the brain to the piano, we begin to understand what tune individual keys produce, but we are not able to comprehend how all keys together create a sonata. Part of the reason is that we are limited with methods to study the whole human brain. We can investigate how individual cells behave and measure electric currents in single brain cells. In animals, we can activate specific parts of the brain and see what effect such activation has. In humans, we are much more limited. We cannot probe a live human brain, and studying a dead brain gives us only limited information.

Nowadays, however, we have some methods that can give us a glimpse of what is happening inside our heads when we perform specific tasks. Two such methods are functional magnetic resonance imaging (fMRI) and positron emission tomography (PET). With functional magnetic resonance imaging, we can monitor blood flow through the brain at any particular moment. We are presuming that active parts of the brain need more oxygen and energy to perform their functions. Brain cells cannot store the energy as muscles do. Therefore, they need a constant supply of both oxygen and sugar (in the form of glucose), which provides power.

Like our cars get the gasoline to drive us around through the tubes leading from the gas tank to the engine, the brain receives the energy and oxygen through the blood vessels, filled with blood. We can imagine the brain as being made up of hundreds or thousands of diminutive engines that are not all active at the same time, but when the need arises. Therefore, these little engines also do not need the same supply of gasoline all the time. Engines working harder need more gas (just as cars use more gasoline when we drive fast), while idling engines need very little petrol. Powerful magnets in the fMRI machines can detect gasoline/blood supply into different parts of the brain. This detection of blood provides a clue to what parts of the brain are active when we perform specific tasks, as they are receiving more blood in a particular moment.

With positron emission tomography, doctors inject a specific compound labeled with weak, harmless radioactivity into our bodies. A machine can then detect the location of this compound. If we inject into the blood radioactive glucose, we can see which part of the brain receives more glucose during a specific action. We presume that this part of the brain needs more energy at that particular time, as it is active in controlling specific activity we perform.

Sex and love always played an important role in human societies and were sparking scientific interests. When Antonie van Leeuwenhoek assembled the first microscope in the late seventeenth century, one of the first things he looked at under his microscope was the human semen (presumably his). As a result of his curiosity, he is accredited with the discovery of sperm. When we learned how to use fMRI and PET to look inside our brains, some scientists were naturally intrigued by what is happening in our brain when we are in love or even when we have sex.

There are many different psychological or mental states when we think of sex, make love, or experience orgasm. Similarly, there are many different states of mind when we are in love. Love can be lust, romantic love, passionate love, attachment, and compassionate love. Therefore, it is challenging to pinpoint brain activity to one particular action. Still, peeking into the brain with the fMRI method gave us some insight into what is happening inside our brains during sex and love.

When we anticipate or imagine experiencing sexual intercourse and its pleasure, we might feel impatient, but we also often feel good even before the actual sex. We feel good because just the anticipation of sex activates parts of the reward system, giving us a pleasant feeling of being fulfilled. fMRI studies confirmed that and showed that during thinking of sex, various parts of the reward system are active.

One such area is the ventral striatum, together with the nucleus accumbens, a region with a central role in the functioning of the reward system. Nucleus accumbens is the main area receiving dopamine signals from the cells in the ventral tegmental area, the primary source of dopamine. Signals are then transmitted to several higher brain areas in the cerebral cortex. One of the areas connected with the ventral striatum is the orbitofrontal cortex. This region plays a vital role in anticipating a reward, evaluating how large the reward will be, and assessing the probability of receiving the reward. Therefore, it is understandable why this area is active when we anticipate a pleasurable affair such as sex.

Another higher brain area activated during anticipation or desire is called a superior parietal cortex, which is located on the top of our brain. It has a function in mental imagining. Usually, when we think of sex, we imagine the mental pictures of either our partner or somebody else sexually attractive. The superior parietal cortex creates these mental pictures.

The amygdala, with its role in processing emotions, and hippocampus, with memories, are two other obvious candidates to be also active when we think of sex. Thinking of sex undoubtedly involves emotions but also needs to activate the memories of previous experiences. Indeed, the amygdala and hippocampus have strong connections with the ventral striatum, and both areas light up in fMRI images of our brain when we think of sex.

Similar brain regions are active when we actually experience sexual pleasure, with one crucial difference. When we have sex, and not just think about it, one additional area, the hypothalamus, is also active. The hypothalamus is part of the ancient brain system and is the primary regulator of all instinctive functions of our bodies. As such, the hypothalamus controls the functioning of sexual organs and sexual instincts in animals. Therefore, it is understandable that the hypothalamus is active during sexual fulfillment. Sex is a biological manifestation of a reproductive function, even though sex is no longer strictly coupled with reproduction in humans.

Insula and anterior cingulate cortex are two other areas that are active in both sex and love. The anterior cingulate cortex is an area that is located deep inside our cerebral cortex, on the top of the limbic system, the primitive mammalian brain. It has close connections with other parts of the cerebral cortex and regions of the limbic system, such as the ventral striatum and amygdala. The roles of the anterior cingulate cortex are not completely clear yet. It is active during many actions we perform. It has some functions in fear and perception of pain. It is connected with the reward system, but most often, the anterior cingulate cortex is associated with resolving conflicting situations and errors.

We want to gather some sweet honey, but stinging bees protect the honey. When hiking, we see an exciting cave that we would like to explore, but we do not know if it is perhaps a bear's den and, therefore, dangerous. In both cases, the anterior cingulate cortex will be activated. It will try to resolve the problem and decide whether the honey is worth the risk and whether our curiosity is stronger than our fear.

A simple task that activates the anterior cingulate cortex is called the Stroop test. In this test, we look at the words meaning various colors, but the letters are written in different colors from the meaning of the word. For example, the word blue is written in yellow letters, the word red in green letters, and the word yellow in red letters. When asked to name the color of the letters, our brains are struggling to name the color. Parts of the brain that read the letters tell us we have to utter the word red, but we look at the green-colored letters. This confusion is called a cognitive interference and causes a delay in our response. The brain has to spend time resolving the conflict between the meaning of the word and the color of the letters. During such tasks, the anterior cingulate cortex is very active.

The anterior cingulate cortex also has a role in regulating our emotions. Our intuitive, reptilian brain is all the time under the control of our highly developed cerebral cortex. Even when the amygdala triggers a strong emotional response, the prefrontal cortex always controls these emotions. Our conscious brain sends the signals through the anterior cingulate cortex to blunt the emotions. This interaction between the prefrontal cortex, anterior cingulate cortex, and amygdala has the responsibility to prevent us from crying, or screaming, or perhaps exposing ourselves in inappropriate circumstances.

So why is this area active when we anticipate or think about sex? There are several possibilities. The anterior cingulate cortex has some role in the functioning of the reward system, so it is perhaps activated just as a part of the dopamine system, expecting pleasurable feelings. Or maybe, it is only activated to resolve the conflicting emotional situation during the fMRI experiment.

For most of us, sex is an intimate affair. We do not want to show sexual arousal to other people besides our intimate partners. When the fMRI scanner screens volunteers' brains, volunteers are alone in the room, yet they are aware that somebody is watching their brains. So, this is an emotionally very conflicting situation. The external stimuli sexually arouse us, but we are aware that somebody is watching us, even if only our brain. This perception creates an emotionally conflicting situation that could activate the anterior cingulate cortex. Perhaps activation of these areas, seen on fMRI scans, is not directly connected with sexual pleasures. The third option is that the anterior cingulate cortex is activated to keep our emotions in check, controlling our responses to the emotional stimuli arising from the more primitive parts of the brain.

These three possibilities why the anterior cingulate cortex is active when we think of sex nicely show how complex are the studies of the human brain. At any time when we are awake, our brains perform multiple functions. Even when resting, we always contemplate something in our minds. During thinking different thoughts, different parts of the brain are active. Therefore, it is difficult to pinpoint a particular action exactly to the activity of the specific brain area that lights up on the fMRI scan. Both fMRI and PET scans are therefore providing us with some suggestions about what areas of the brain might be involved in the regulation of specific activity. Still, they do not give us exact answers to difficult questions about which parts of the brain regulate certain behaviors.

Insula, meaning island in Latin, is one of the least understood parts of the brain, partially due to its location deep inside the cerebral cortex. It has an important role in most sensory perceptions, including smell, taste, hearing, balance, and sensory perceptions from skin and muscles. Insula is also activated in pain and helps other parts of the brain in controlling the muscles. A section of the insula is involved in the cognitive functions, but one portion is also active during emotions and in the social context. fMRI scans showed that almost any emotional trigger such as disgust, fear, happiness, sadness, and sexually arousing pictures activates the insula.

Insula probably has a major role in emotional experiences and in subjective feelings toward certain emotions. Insula also has an important function in empathy. When we feel empathic, the insula lights up on fMRI scans. People with damaged insula express lower levels of empathy. However, it might not be that such people are unable to feel empathy. Instead, they probably cannot recognize emotions such as fear, sadness, and happiness on other people's faces and cannot respond empathically. So, as sex and love are associated with strong emotions, it is hardly surprising that the insula is activated in both sex and love, even though we do not know what the insula actually does. Interestingly, the insula is also one of the few brain regions that are activated during hatred.

Sex is not equal to love, but the difference seems to be relatively small when it comes to brain activity. fMRI scans of people being madly in love or living in harmonious, lovable relationships for 20 odd years showed that similar brain areas are active in both sex and love. In love, as in sex, we can see robust activation of the reward circuitry, although perhaps with slight differences. While in both anticipation of sex and during sexual arousal there is a strong activation of the ventral striatum and nucleus accumbens, people in love show more activity in the nearby caudate nucleus, which is like nucleus accumbens receiver of dopamine signals.

The ventral tegmental area, a source of dopamine, a "feeling good" hormone, is activated when we are in love or experience sexual pleasure. Interestingly, this area is more active when we are in love. This difference suggests that love is a more potent drug than sex and that the "high" we get from being in love might be stronger than the "high" we get from sex. This is probably true as even though sex is a very gratifying experience, it is usually short-lasting, while the feelings of being in love last for weeks, months, or even years.

The hypothalamus, the site of the autonomous brain, is active in both love and sex. While it is clear why the hypothalamus, with its centers directly regulating the function of the reproductive organs, is active during sexual intimacy, it might be less obvious what the hypothalamus has with love. Love is an autonomic response that we cannot consciously control, and parts of the hypothalamus are involved in the regulation of social contacts in animals. Although clearly sex and love are not compulsory intertwined, sexual intimacy is an integral part of any love. Even casual sex can develop into love. Therefore, there are two explanations for the involvement of the hypothalamus in love. Perhaps the hypothalamus is activated as a part of the autonomic regulation of our social instincts, or the hypothalamus is active in anticipation of sexual intimacy and, eventually, reproduction, which is usually an important aspect of love.

Sex and love also activate the insula and anterior cingulate cortex, but there are subtle differences in the precise location of the activity in these two brain regions. These variations suggest that areas active in love and sex might perform similar functions, but in a different context. The anterior cingulate cortex has a role in resolving conflicting situations. When looking at brain activation during sexual arousal in volunteers, this area might be activated due to emotional distress caused by demonstrating sexual pleasure during brain scans in a public or semi-public place. In love, the anterior cingulate cortex might have a more critical role. When we fall in love, we get close to a stranger, but this is against our deeply embedded instincts.

Our natural drives, evolved during millions of years, are to be afraid of strangers. Outsiders usually meant trouble for our ancestors. They wanted our food, our shelters, or even our partners. Therefore, we learned to avoid strangers and be hostile toward aliens. We hold it in our primitive brain to be unfriendly or afraid of everybody that does not belong to our inner circle. Although thousands of years of civilization somewhat lessened the fear of strangers, this basic instinct is still encoded in our genes. Unfortunately, we can still see and experience this in everyday life. Even in the most advanced societies, racial prejudices, racial hatred, and even armed conflicts are still widespread. This fear of strangers is also a problem for love and relationship. We have to overcome our basic instincts and allow an outsider close to our hearts. Anterior cingulate cortex might be, therefore, activated in love to resolve this internal conflict.

Activation of similar brain areas when we are in love or just crave sexual pleasure does not suggest that sex and love are equal. The fMRI and PET methods are very crude methods that cannot give specific answers when we peek into the brain. They offer us only a glimpse of what is happening inside our heads. Tens of billions of cells form our brain, and each part of the brain has millions of cells. Activation of one part of the brain, seen on the fMRI scan, does not necessarily mean that the same cells were activated by two different stimuli (love and sex), as neither fMRI nor PET could distinguish between individual cells. It is entirely plausible that different cells within the same brain areas are active, but we cannot detect these differences with methods such as fMRI and PET.

Brain activity during love and lust points to the activation of the general brain circuitry involved in sex and love. But the methods we use today are not precise enough to tell us in detail what is going on when we are in love or when we have sex. Undoubtedly, there are subtle differences in the brain activity between lust, romantic love, compassionate love, or casual sex that we cannot discern just yet. Further supporting the indication that areas activated in love and sex are only general brain circuits activated in gratifying experiences is that the same regions are activated in other pleasurable experiences such as eating good food or taking drugs. All these suggest that we identified only the activity of a very general circuitry that governs pleasant feelings and craving for satisfaction. Unfortunately, we do not have yet an insight into precise mechanisms that are going on in our brain when we are in love or when we experience the pleasure in sex.

We know that both sex and love are pleasing experiences. Therefore, it is understandable why the areas associated with "feeling good" and other emotional parts of the brain are active when we feel emotions of love, either romantic or compassionate love. These insights into brain activity, highlighting the importance of the reward system in the feelings of love, suggest that love is less an emotion and more motivation to stay together with our partner. But how specific brain areas modulate and distinguish between feelings of lust, sexual desire, romantic love, attachment, and compassionate, friendly love is unclear. We do not have tools at present to study these emotions or motivations at the cellular levels. Unlike some other workings of the brain such as controlling muscles; detecting sounds, sights, smells, or tastes; or even craving for drugs, which we can study in animals, we cannot explore love in animal brains as there is no real equivalent to the romantic human love in animals.

However, let's look at the prairie voles again. We can see that several areas active in the human brain on love are also involved in the formation of attachment between male and female prairie vole. Dopamine and the reward system have a central role in the bonding between voles, but oxytocin receptors are also present in the voles' cingulate cortex and insula. The presence of oxytocin receptors in the cingulate cortex and insula suggests that these two areas are indeed crucial in the attachment and bonding between partners. Although we do not know the precise roles of the cingulate cortex and insula in love and attachment, it looks like they have a long evolutionary history of being involved in the regulation of partnerships.

One interesting aspect of brain activity during love is which brain areas are less active than usual. One such region is the amygdala, which has important roles in processing emotions, especially fear. When we are scared of something, the amygdala activates our stress system. We start to sweat, our heartbeats speed up, and we begin to breathe quickly. When we are in love and away from our partner, we often experience some of these symptoms.

In contrast, when we are close to our partners, we usually feel safe and relaxed. The reduced activity of the amygdala sparks these calm feelings. Why is amygdala deactivated is not entirely understood. One explanation suggests that lessening the fear is essential for getting close to a stranger we suddenly love. While the anterior cingulate cortex analyzes the conflicting situation, whether it is safe to be close to a stranger, low activity of the amygdala allows us to feel relaxed and content when in the company of our new partner, with whom we fell in love.

Other higher brain areas partially deactivated when we are madly in love are parts of the frontal cortex, especially the prefrontal cortex. This brain area is involved, among others, in the judgment and evaluation of other people and situations. As we all know, love is blind. And this might not be only the ancient saying. Deactivation of the frontal cortex seems to be why people in love are often unable to judge their lover's character honestly.

The prefrontal cortex, part of the frontal cortex, is involved in mentalizing. It is the main area for higher cognitive functions such as thinking, reasoning, anticipating, and evaluating other people's emotions and intentions. We all know these functions are reduced when we are madly in love. This reduction is due to the lower activity of the prefrontal cortex. However, the frontal cortex and especially prefrontal cortex are less active only during the initial stages of love, when we are madly in love. When love progresses to attachment and compassionate love, both the prefrontal cortex and frontal cortex are operating normally again.

So, love is blind, but only temporarily, which is why many relationships do not endure beyond the first phase of romantic love. When we are madly in love, and our frontal cortex is partially shut down, we see our partners how we want to see them, not how they really are. We establish an idealistic mental picture of a partner that might not always overlap with the truth. We see and amplify in our brain the virtuous characteristics of our spouse, but everybody has some good and some bad sides. When the first rush of love wears off, and the frontal cortex starts to function fully again, we are suddenly confronted with the reality of our partner, and many relationships break up. But if we endure, if we find more positive than negative emotions even with the frontal cortex up and running, our relationship will survive. Only then can the relationship develop into lasting, lovable unison of two people.

Besides sex and romantic love, maternal love is a deep and ancient emotion present in all mammals. When we look at the mothers' brains, we see many parallels in which brain areas are active in love, sex, and maternal love. But there are two crucial differences. One difference is that in motherly love, the hypothalamus is not as active as in the sexual context or during romantic love.

The hypothalamus is the center of instinctive behaviors, including sexual drives and sexual behavior. As romantic love is associated with reproduction, it is understandable why the hypothalamus is also active when we are in love. On the other hand, maternal love has nothing to do with sex. This explains why the hypothalamus is not as active when mothers look at the photos of their babies as it is when we think of sex. The medial preoptic area, one of the most important regions triggering and regulating maternal attachment and behavior, is part of the hypothalamus. But it is a small part of the hypothalamus, and it might be primarily a trigger of maternal behavior, not necessarily needed to sustain maternal instinct. Therefore, the preoptic area might be too small, or just not active enough, to be detected as strongly operational with fMRI or PET brain scans of maternal brains.

One area of the brain is activated in mothers but not in lovers. This area is an area involved in recognition of faces. When mothers look at the photographs of their infants, the facial recognition area is strongly activated. But when lovers look at photos of their soulmates, this same area remains switched off. This difference might reflect the necessity for mothers to recognize the facial expressions of their babies. Mothers have to recognize whether the baby is hungry, sleepy, happy, or perhaps in pain. Facial expressions in young babies are not as strong and not as apparent as in adult humans. Therefore, an active facial recognition area might be necessary for mothers to recognize the babies' needs and provide proper care. With lovers, however, we usually quickly remember their faces. They embed firmly into our memory. Although recognizing our lovers' facial expressions is an important part of being in love, it is perhaps not as tricky as it is for mothers to discern the facial expressions of young babies.

All our feelings come from the brain, and love is no exception. We have learned what parts of the brain are active when we are in love. But with the current methods that allow us to peek inside our brains, we can only identify broad circuits and not specific workings of our minds. We can see how pressing several keys on the piano produces one chord, but we do not understand how multiple chords together form a musical masterpiece. Just as the same chords create different music when joined together in a different sequence, subtle differences in the activation of the same brain regions produce different responses such as satiety, sexual fulfillment, or feelings of love and attachment.

We know that love is a rewarding experience, and that is why it is addictive. We also understand why love is blind and why we feel content in the embrace of our intimate partner. We know that both intuitive and conscious parts of the brain work together in love. But what is really going on inside our brains when we are in love? Why do we fall in love with one person but not another? Why some people love somebody their whole life, while others end up in many broken relationships? Why are some faithful to their partners, and others always look for a greener pasture? All these are unanswered questions. We will need better methods to look inside our brains and a better understanding of how our brains work in their entirety before we will understand what is happening in our brains during passion and love.

# Chapter 23
# Man and Woman Madly in Love

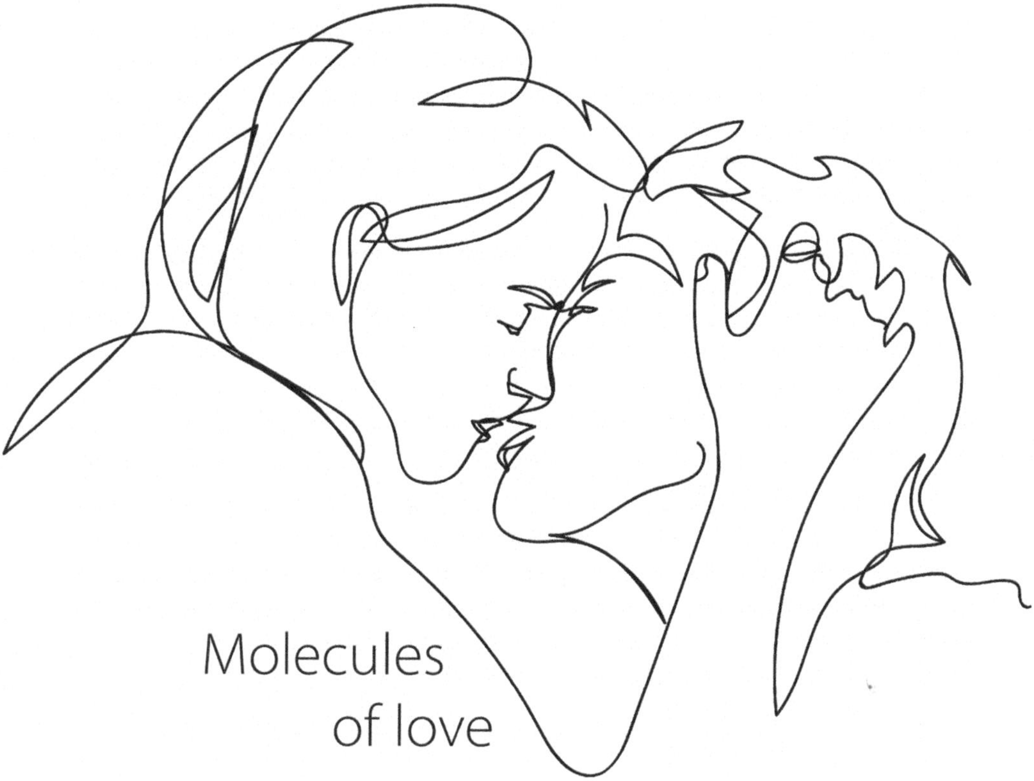


*"It has made me better loving you ... it has made me wiser, and easier, and brighter. I used to want a great many things before, and to be angry that I did not have them. Theoretically, I was satisfied. I flattered myself that I had limited my wants. But I was subject to irritation; I used to have morbid sterile hateful fits of hunger, of desire. Now I really am satisfied, because I can't think of anything better,"* said Osmond to Isabel in Henry James's The Portrait of a Lady. Clearly, Osmond talks about mature love, long-lasting attachment between two people and not about the crazy romantic love at the beginning of the relationships.

Hormones and neurotransmitters, chemical messengers in our body and brain, play essential roles in every aspect of our lives. Hormones and neurotransmitters tell us when we are hungry or thirsty. They regulate our fear and dread. Hormones make us



G. Majdic, *Soul Mate Biology*, Fascinating Life Sciences,
https://doi.org/10.1007/978-3-030-67212-6_23

shiver when we are exposed to cold; hormones regulate how strong are our bones. Hormones and neurotransmitters tell us when to fight and when to flee to preserve our lives. Therefore, it is not surprising that hormones and neurotransmitters have essential roles in both lust and love.

Hormones, most intimately connected with reproduction and sex, but not necessarily with love, are sex hormones, androgens, and estrogens. Both are present in both sexes, but with different roles. Men have much more androgens than estrogens, while in women, estrogens are more important than androgens. Yet, both androgens in women and estrogens in men have essential physiological functions. In women, androgens are responsible for the growth of pubic hair, and if in excess, they trigger the growth of the facial and back hair and cause a hoarse voice. In men, estrogens are essential for bone growth, but they can trigger breasts' appearance in men when in excess.

Testosterone, the primary male sex hormone, is responsible for the development of male sexual organs and male characteristics such as deep voice and facial hair. It is essential for the male libido, and stimulates aggressive behavior. Estradiol, the primary female sex hormone, is responsible for the growth of breasts; thicker, smoother skin; and changes in the internal reproductive organs in anticipation of pregnancy during every menstrual cycle. Estradiol undoubtedly also affects behavior, although these effects are not well explored yet in women.

In animals, estradiol triggers the sexual instinct in females. It activates the brain circuits that regulate behavior leading to copulation. Estradiol also makes females receptive to the advances of males. Estradiol released from the ovary is transported by the blood to the brain. In the hypothalamus, estradiol, together with external stimuli from male, triggers the lordosis, allowing the coitus. Testosterone has similar effects. It regulates sexual desire, and male animals without testosterone lose the libido. In some animals such as mice, adding testosterone during adult life even triggers male-typical mating behavior in females. Female mice treated with testosterone try to have sexual intercourse with other females in a cage, even though they are perfectly normal females.

In humans, sexual instincts are unquestionably more complicated than in animals. In addition to androgens and estrogens, emotional and psychological factors have crucial roles in sexual desire. Furthermore, we do not have sex only around ovulation, as most animals do, confirming that in women, sexual desire is not exclusively dependent on the presence of estrogens. Undoubtedly, estradiol also affects the female libido, although it is not yet understood how exactly and what are all the interactions with the other brain systems. For the last 20 years, there were suggestions that testosterone is essential for the female libido, but this is far from conclusive. Many studies failed to show the positive effects of testosterone on the female libido. Nevertheless, testosterone is often prescribed to women with low sexual desire, but the effects of such treatments are very doubtful.

Testosterone and estradiol undeniably have roles in the feelings of lust, as this is the evolutionary most primitive instinct, forcing us to reproduce and transfer our genes into the next generations. But lust is not love, and it seems that sex hormones agree with this. At the beginning of a new romantic relationship, testosterone levels in men fall, while they increase slightly in women. One reason for the fall in testosterone levels in men might be the commitment. As testosterone stimulates male libido, it could drive a male to seek more sexual intercourses with different partners. This enhances the chance his genes will pass onto the next generations. But once we are on a path toward commitment, this is undesirable. Our species evolved that males should stay around one partner to protect and care for both the mother and a child. Although sexual activity is an integral part of the loveable relationship, moderate testosterone levels present in males in love are more than enough to sustain the male libido. Therefore, there is no need to have excess testosterone when we are on the path to commit ourselves to one partner.

Why is testosterone increased in women at the beginning of love is less obvious. Higher natural levels of testosterone are connected with more assertive behavior in women. Perhaps this is part of the reason testosterone levels increase. Women need to be self-confident when entering a long-term relationship, as stakes for females when entering a relationship with the prospect of creating a family are much higher than for men.

Male animals or human gentlemen can easily walk away from a relationship and seek greener pastures elsewhere, even if they have fathered a child. Risks for women are higher when entering the relationship. A woman must be sure that she met the right partner for a long-term relationship, the one that will help her raise the children. This certainly requires assertiveness. Interestingly, after 12–24 months of being in love, testosterone levels rise in men and fall in women, returning to the average levels for all men and women. Why it happens is also unknown, but we can speculate that our commitment has grown strong after 1–2 years in a loyal, loveable relationship. Women no longer need extra assertiveness, and men do not need a reduced sex drive, as they have already committed themselves to the one partner.

Although testosterone and estradiol undoubtedly have a significant role in lust and sexual desire, their role in creating a loveable relationship is probably marginal. Both hormones might facilitate the first encounters through longing for sex, but they do not trigger the feelings of romantic love, yet alone attachment. High sexual drive is actually quite often inversely related to staying in a committed, long-lasting relationship. In love and attachment, other hormones and neurotransmitters have more prominent roles.

One of the main factors involved in romantic love is dopamine, the principal neurotransmitter in the reward system and an essential part of "feel good" emotions. Dopamine is mainly produced in the brain region called the ventral tegmental area. Dopamine from the ventral tegmental area travels to and excites the nucleus accumbens. Through the nucleus accumbens, it reaches the anterior cingulate cortex, orbitofrontal cortex, amygdala, prefrontal cortex, and insula, all brain areas that are activated or deactivated when we are in love. All these areas contain abundant dopamine receptors and are therefore responsive to dopamine.

Although we do not have any direct proofs that dopamine is essential for experiencing romantic love, love is undoubtedly similar to addiction. In the early stages of passionate, romantic love, men and women show several behavioral patterns akin to addiction. When in love, we crave for the beloved, and all our life is focused around our darling. We feel a rush of exhilaration when we are together with our sweetheart. We experience withdrawal symptoms when we are separated. Withdrawal symptoms after breaking up from a romantic relationship can be remarkably similar to the withdrawal from drugs. Symptoms include anxiety, insomnia, obsessive thinking, lethargy, irritability, and others.

Rejected lovers often go to the extremes, doing degrading or dangerous things to win back the heart of the beloved. Like with drugs, relapse is always a severe problem, and unfulfilled love can also trigger a relapse of romantic feelings even years after breaking up. After many years, seeing our unfulfilled love can switch on the memories of being in love and renew the craving for that person. Dopamine has essential roles also in the formation of pair bonds in animals. All these suggest that romantic love is associated with the activation of the reward system in our brains.

Precisely what is happening with our dopamine when we fall in love is still a secret. During the early stages of being in love, love triggers dopamine release from the ventral tegmental area. This dopamine activates parts of the brain that are responsible for the euphoric feeling when we are with our partner and for withdrawal symptoms when we are separated. Love is, therefore, some kind of addiction. In the early stages of love, being close to the person we fell for triggers "feel good" emotions. Essentially, we experience being "high." As this is a rewarding feeling, we become addicted to it. We want to stay close to our partner because we feel blessed in its vicinity. This addiction tricks us into the long-term relationship, at least long enough so the blokes help moms raise the children.

Romantic love is undoubtedly a great feeling, but it is often also very stressful. We are constantly worried about whether we have chosen the right partner. We are afraid the beloved one does not share our feelings. We are stressed when we are separated from our sweetheart, and not the least, we have to overcome our natural fear of strangers. Therefore, it is not surprising that levels of stress hormones, adrenaline and cortisol, are elevated during the early stages of romantic love. We do not understand precisely the role of stress in love, beyond the fact that being madly in love is both a pleasant and stressful experience.

The primary purpose of adrenaline and cortisol in our system is to prepare the body to fight or flight. When in danger, we have to fight for survival or flee from the threat. Adrenaline and cortisol achieve this through various mechanisms. They speed up our heartbeat, deepen our breathing, and prepare the muscles to act. Indeed, when we are in love, our hearts beat faster, and our knees are often weak, like when we are mortified. We do not have to fight or flee from the danger when we are in love (unless we have an affair with a married person), but we are worried about our feelings, and we are, therefore, under stress. Cortisol also makes us more alert and less sleepy. When we are madly in love, we often cannot sleep as we keep thinking about our beloved. As Dr. Seuss said: "You know you're in love when you can't fall asleep because reality is finally better than your dreams." Less poetically, part of the reason why we cannot sleep is alertness caused by cortisol.

Although in the early stages of romantic love cortisol levels are high, partners living in a long-lasting, harmonious relationship have lower levels of stress hormones than single adults. Initial falling in love is a stressful experience, but being in a lasting lovable relationship is beneficial for our stress levels. We feel calm and content, and there is no need to be worried. So, levels of stress hormones drop, and this is likely connected with the general health benefits of living in an amicable relationship.

Cortisol affects the activity of another important neurotransmitter in the brain, serotonin. Serotonin has many essential and complicated roles in our brains, most of which we do not yet understand. Although there is only 1 form of serotonin, there are at least 15 specific serotonin receptors. Through these receptors, serotonin achieves many different effects in our brains. These receptors are present in various parts of the brain, and they produce different, sometimes even controversial, results.

In the early stages of love, when we are trapped in romantic love, serotonin levels are low, most likely influenced by increased cortisol levels. We still do not understand the complicated actions of serotonin in the brain, but low levels of serotonin are associated with two common psychiatric disorders, major depressive disorder and obsessive-compulsive disorder. The latter is characterized by typical obsessive thoughts and compulsory behavior. People suffering from this severe disease have obsessive thoughts, urges, or mental images. These obsessive thoughts often cause anxiety. They also lead to compulsory behaviors such as incessant cleaning, ordering and arranging items in particular patterns, or checking if doors are locked and stoves or switched off. After a bout of compulsory behavior, patients might experience a short relief. Unfortunately, this usually does not last long, and a new cycle of obsessive thoughts and compulsory behavior starts over and over again.

People madly in love often behave quite similarly. We have obsessive thoughts about the beloved. We relentlessly think about our loved one. We feel anxious when separated, wondering if the person we fell in love with shares our feelings. We anxiously wait for the next rendezvous, which often leads to compulsory behavior. We repeatedly take a cellphone out of the pocket and check if any messages have arrived or we keep calling our new dearest. Once we are with our beloved, we feel temporary relief. Soon, however, we are trapped in another circle of temporary madness when we are separated from our loved one. So, in essence, our behavior when we are in love is similar to obsessive-compulsive disorder. But despite similarities, love is certainly not a psychiatric disease. It is unfair to the patients with the obsessive compulsory disorder to compare love to this disease. The obsessive compulsory disorder is a severe and life-changing illness. Although love often changes our lives, it is certainly not affecting us as severely as the obsessive compulsory disorder affects the patients' lives.

Fortunately, the symptoms of madness in love disappear after a while. We pass through the initial stages of the romantic relationship. When our love matures, serotonin levels rise back to the normal levels, and we feel calm and content. If the early relationship progress to long-lasting love, we develop an attachment to the person we love, but we lose obsessive thoughts. We become used to living together with our partner, so we do not have to show compulsory behavior any longer. We know and trust our partner that they will be at home when we come from work, so there is no need to continually check our cellphones. We still send each other messages and call each other when we are separated, but for different reasons. We do not feel stressed and worried. Instead, we miss the closeness of our partner, the safety we feel when we are together, and the gratification we sense when in the embrace of our dearest.

Hormones and neurotransmitters made love stressful, addictive, and obsessive. But love is also closeness, safety, affection, care, companionship, compassion, devotion, endearment, intimacy, and many other feelings. Most of these are caused by two other hormones. Although these two other hormones are intimately coupled to dopamine and stress, they are the hormones of bonding and attachment. They are hormones responsible for the feelings of contentedness, trust, and closeness. You guessed—we are now talking about oxytocin and vasopressin.

Oxytocin and vasopressin are the main hormones responsible for the lifelong bonds between male and female prairie voles. Oxytocin is essential for triggering the maternal instincts in mammals and is important during labor by stimulating womb contractions. Although these functions are mostly, if not exclusively, associated with mammals and to some extent with birds, oxytocin and vasopressin are ancient molecules. Genes similar to oxytocin and vasopressin are at least 600 million years old. They can be found today in ringworms, shellfish, crustaceans, and many other animals. However, only vertebrates have two separate genes, oxytocin and vasopressin. In contrast, other animals have one common predecessor to these two peptides.

Vasopressin and oxytocin were formed by the duplication of an ancient gene in early vertebrates. Slowly through evolution, they acquired different functions in different groups of animals. In earthworms, the common predecessor regulates egg-laying behavior. In cuttlefish, it is involved in memory formation. In locusts, it regulates bodily fluids. In some lizards and toads, these two hormones regulate courtship behavior. They stimulate flocking in zebra finches and vocal acoustics in some fish. In mammals, including humans, oxytocin's primary role is triggering womb contractions during the labor and expulsion of milk during nursing. The primary function of vasopressin is the regulation of the water balance in the body. It reduces the secretion of fluids when we are dehydrated and consequently also regulates the blood pressure. These are all hormonal functions of vasopressin and oxytocin. However, both hormones act as neurotransmitters in the brain and have essential roles in regulating social behavior.

Oxytocin and vasopressin are chemically very similar molecules. They are both made from nine amino acids, and only two of these building blocks are different. Consequently, their action in the brain is quite similar. Although each has its own receptor, oxytocin can, when present in large amounts, activate the vasopressin receptor, and vasopressin can similarly stimulate the oxytocin receptor. Therefore, it is difficult to completely distinguish their individual roles in the brain.

As a general rule, oxytocin seems to be more involved in the feelings of intimacy, and vasopressin has important roles in caring for and defending our love nest. Oxytocin is more connected with the passive aspect of love. Vasopressin is more important for the active elements of the partnership, such as shielding our sweetheart from potential competitors and actively defending our dearest partner and children from peril. Furthermore, at least in the prairie voles, oxytocin is more important for pair bonding in females, while vasopressin is more vital for attachment in males. Although we do not know if this difference between sexes is also present in humans, it fits well with the passive and active aspects of love being regulated separately by oxytocin and vasopressin. Women are usually more prone to cuddling and enjoy intimacy. While you should never cross a path of the mother protecting her child, it is usually males who take a more active role in caring for and protecting their love nests from danger.

We do not yet understand all the functions of oxytocin and vasopressin. We know they have crucial roles in regulating the pair bonding in voles and some other animals such as marmoset monkeys. They are the master regulators of all behavioral responses that lead to the attachment between animal males and females. They regulate mutual care for the offspring, aggressive behavior toward intruders, and grief after the loss of the partner in monogamous animals.

Roles of oxytocin and vasopressin in the human brain are much more challenging to study. Yet, we have some evidence that both have a role in regulating our social behavior. As in animals, oxytocin is essential in human labor and during nursing. It likely has similar effects on the mother-child bond as in animals. Adding oxytocin increases positive feelings toward children in both men and women, and variations in the gene for oxytocin receptor correlate with the strength of the maternal attachment. Moms and dads who show more affection and play more with their children usually have higher oxytocin levels in their bodies. Also, women that are not mothers are more attentive to children when they are exposed to oxytocin.

The role of vasopressin in modulating parental behavior in humans is not yet known. However, both vasopressin and oxytocin are promoting parental care in many animals. In line with the vasopressin's active role in animals, vasopressin in humans likely increases aggressive defensive behavior. Unlike testosterone, which stimulates aggressive behavior in general, vasopressin does not make us more aggressive in any situation. However, if we have to defend something we possess, and particularly someone dear to our heart, vasopressin enhances our aggressive response. We gather strength and courage to fight the intruders wanting to take away something that we care for.

In humans, both vasopressin and oxytocin are released in our bodies during sex. Vasopressin is elevated in males during sexual arousal. Nursing a baby triggers oxytocin release so that milk can flow from the breasts. Likewise, caressing nipples during intimacy stimulates oxytocin release in women. Even hugs and kisses release some oxytocin. In both men and women, oxytocin is flooding our brains and bodies during orgasm. At least some of the pleasant sensations during orgasm are a consequence of oxytocin. Oxytocin during orgasm has a role in both sexes. It helps to eject the semen in males and stimulates contractions of the womb in females. At the same time, oxytocin also acts in our brains. It triggers feelings of closeness, attachment, and desire to be together.

Oxytocin and vasopressin activate dopamine release in the ventral tegmental area. They also modulate the response to dopamine in various parts of the brain that are receptive to the dopamine. Both oxytocin and vasopressin activate the amygdala, although differently. While oxytocin reduces fear and anxiety, vasopressin increases alertness and apprehension. Again, this is connected with the presumed diverse roles of both hormones: oxytocin a cuddly molecule and vasopressin a defensively aggressive hormone. Receptors for both vasopressin and oxytocin are also present in the hippocampus. This is likely connected with remembering both our partners and pleasant feelings when we are together.

Craving to spend time together with our beloved is an integral part of love. It is triggered by oxytocin, which causes an overflow of dopamine in our brains. Feeling relaxed and content in our partner's presence is another crucial aspect of love, accomplished by oxytocin through the regulation of the amygdala. The third essential segment of any lovable relationship is trust between partners, and oxytocin plays a vital role in regulating trust between people.

With increased levels of oxytocin, we trust other people more, either our lovers or only a bank clerk, trying to convince us to buy particular shares. Partners are more agreeable under the influence of oxytocin. People are willing to invest in more risky schemes offered by a stockbroker when they are exposed to the oxytocin. However, it seems that personal contact is necessary for increased trust. When people communicate with a bank clerk through the computer, oxytocin does not increase our confidence. Trust, enjoyable feelings, and being relaxed in the presence of our partners are certainly some of the most important fragments of love. All of them are most likely triggered by oxytocin, possibly together with vasopressin.

When it comes to trust, oxytocin also has its dark side. Yes, oxytocin enhances trust and it boosts our "feel good" emotions, but it seems that it does so only toward the people we are familiar with. When we face a stranger, especially if this stranger is very different from us, oxytocin could increase distrust and hate toward the strangers. One interesting psychological study with a hypothetical situation in Holland showed that people were more likely to help a stranger with a name than a nameless person under oxytocin's influence. But there was a glitch. People in the study (all Dutch) were only willing to help if a person had a Dutch name. If a person they could rescue had a German or Arabic name, they were less likely to save this person when under the influence of oxytocin. So, oxytocin may not be a general trusting and cuddling neurotransmitter; it may also be a racist neurotransmitter.

Perhaps having a racist neurotransmitter in our bodies sounds horrendous, but it does have an evolutionary sense. Throughout evolution, we had to be afraid of strangers. Therefore, it makes perfect sense that we have a neurotransmitter that makes us more trustful toward our kin but, at the same time, distrustful toward outsiders. However, this cannot be an excuse for racial hatred and racial prejudices. Even though we might have embedded distrust toward strangers in our genes, we are also striving to be civilized. A critical aspect of being civilized is being respectful to all people. As with many other instincts, our conscious brain is more than capable of surpassing the basic urges of distrust toward those that are not our kin.

Although there might be evolutionary wisdom in oxytocin regulating trust and distrust, it does mean we have to be very careful with oxytocin. In the past, some have suggested that oxytocin could be used as a treatment for antisocial behavior. However, if oxytocin increases the distrust for foreigners, such treatment in antisocial people might have the opposite effect from what we would long for. Instead of making people trustier and more friendly, it might actually make them more aggressive toward strangers. This nicely shows that we have to be very careful when meddling with the hormones and neurotransmitters in our bodies. Many hormones and neurotransmitters in our bodies have numerous functions. Trying to modulate one function of the hormone or neurotransmitter might produce other undesirable effects.

Our brain is actively learning and adapting to different situations throughout our lives. Various early childhood experiences shape our brains permanently. This also includes the oxytocin and vasopressin system. Oxytocin and vasopressin are active in moms and dads, but they are also active in children who respond to parental care. A smile from mom or dad brings up a laugh in a baby. Children feel safe and content in the laps of their parents. This happiness is at least partially caused by the release of oxytocin, and possibly vasopressin, in the young brain.

But this release is not only crucial for a smile and a cuddle in the children; it shapes how the oxytocin and vasopressin will function in our brains throughout our lives. Neglected and abused children often have trouble forming lovable relationships. They fall in love less often. They have more difficulty sustaining the lovable relationships, especially in progressing from romantic love to long-lasting attachment. This is caused by permanent changes in the action of oxytocin and vasopressin, influenced by early neglect and abuse. Montane voles do not form a long-term monogamous bond simply because they do not have vasopressin receptors present in one specific brain area. Neglected and abused children have most likely trouble creating lovable relations as oxytocin and vasopressin receptors in their brains are not functioning correctly.

We have to give our children wings and allow them to spread their wings and soar through life without us when they are old enough. But it is equally important that children grew roots in the compassionate family. We have to offer them care and love when they need it, or else they will bear the consequences for their whole life. Without love and care, they will be unable to experience true love when they grow up. Not just because they will not learn how to love from their parents but because their brains will act differently from the brains of children who grew up in a caring and compassionate family. And this is important, as life without love is like a rose bush without flowers. Yes, it can grow and climb up toward the sky. It can have a strong trunk and prickly thorns. But only the roses give it the meaning and splendor in this beautiful and mysterious world.

# Chapter 24
# What Is Love?

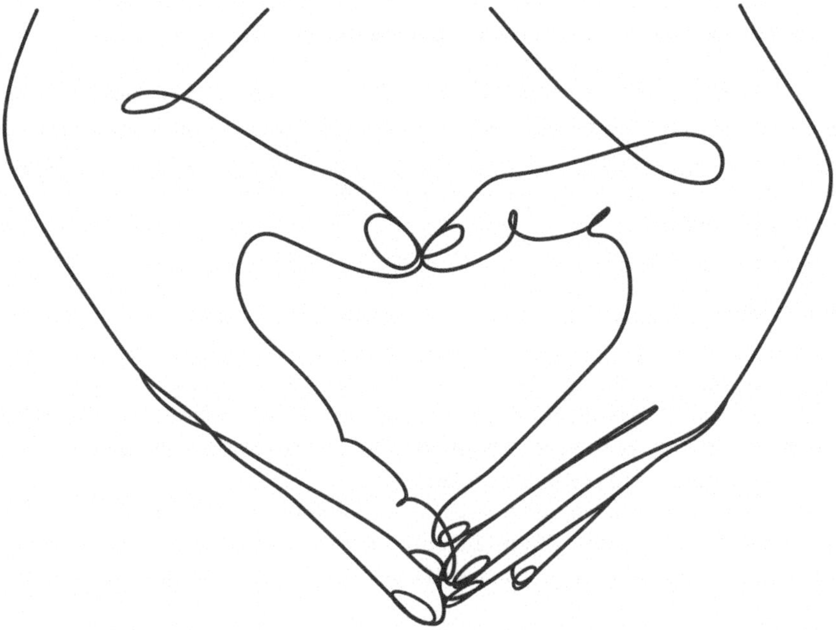

What scientists know about love?

*"When you fall in love, it is a temporary madness. It erupts like an earthquake, and then it subsides. And when it subsides, you have to make a decision. You have to work out whether your roots are to become so entwined together that it is inconceivable that you should ever part. Because this is what love is. Love is not breathlessness, it is not excitement, it is not the desire to mate every second of the day. It is not lying awake at night imagining that he is kissing every part of your body. No … don't blush. I am telling you some truths. For that is just being in love; which any of us can convince ourselves we are. Love itself is what is left over, when being in love has burned away. Doesn't sound very exciting, does it? But it is!"* explained Dr. Iannis to his daughter Pelagia in a Luis De Bernier book Captain Corelli's Mandolin. During the Second World War, Pelagia, living on the Greek island of Kefalonia, falls in love with the Italian officer Antonio Corelli. Love that should not be there, yet it existed in this book, as in many true life stories.



G. Majdic, *Soul Mate Biology*, Fascinating Life Sciences,
https://doi.org/10.1007/978-3-030-67212-6_24

So, what is love? The straightforward question, but nobody knows the answer. If we ask a hundred people, we will receive a hundred different answers. Love is madness, love is closeness, love is harmony, love is family, love is a poem, love is a story. Love is also chemical reactions in our brain, and love is something we dream about. Love is all these and much more. There is no single or simple answer to what love is. As Duke Wellington once said: "Love is indescribable and unconditional. I could tell you a thousand things that it is not, but not one that it is."

As we have learned from animals, hormones, particularly oxytocin and vasopressin, are essential for the feelings of attachment and closeness to another person. Dopamine makes us "high" when we are in love. Testosterone in men and estradiol (perhaps together with testosterone) in women trigger feelings of lust, desire to mate, and desire to reproduce and thus ensure the continuation of our species into the next generation.

The hypothalamus triggers the sexual instinct, a desire to mate. The amygdala reduces our fears of strangers, so we can develop closeness with an outsider. Ventral pallidum and nucleus accumbens trigger "feel-good" emotions when we are with our loved one. The hippocampus remembers pleasant feelings we experienced in the company of our sweetheart. The cingulate cortex helps us resolve a paradoxical situation, a desire to be with a potentially dangerous stranger. Insula combines basic instincts with higher reasoning. The prefrontal cortex temporarily goes backstage until we develop a more permanent relationship.

So how can love be a single feeling? As we all know, it is not. Love is a mixture of different emotions. Some of these feelings are quite different, perhaps even controversial. Love is a storm, a hurricane of emotions in which we are swirling, like a leaf in a tornado.

Love can be separated into three different stages. The first, initial phase starts with being madly in love. This initial stage can begin from the lust, but this is not essential. Indeed, we often begin to love somebody before becoming intimate, but love also works the other way around. We might plan to have casual sex, but one-night stands often lead to love. Attachment hormones are released during sex. Even if we did not plan ahead, we might fall in love with a person we initially planned to have casual sex only. So, there is a problem with the one-night stands. Unless we are on vacation in a foreign country and will never meet a person again, planned casual sex always has a danger of developing into something more.

In this initial stage of lustful love, high passion befalls partners. They share intimacy and long to be together all the time. Partners commit themselves to one another, and all they do is focused on their loved ones. When lovers are not together, they think incessantly about their loved one and cannot wait until they meet again. In this phase, lovers are excited and stressed. Their stress hormones are high, and serotonin levels are low. Although they long for sexual relations, their testosterone levels are down, certainly in men. During this early stage of love, dopamine is flushing our brains. We yearn for the feeling of being "high," an emotion we experience when we are together with our loved one.

In early love, the frontal cortex, our conscious brain, runs in a low gear, and we see roses everywhere. We idolize our partners, and we do not see their faults. We not only avoid any conflicts; we believe that there will never be any. We are convinced that we are a perfect match and will remain inseparable for the rest of our lives. Everything we do in life is so much better when we do it in the company of our beloved. During this stage, most romantic poems are written, and songs about immortal love are sung. This is the madness of love. The craziness our grandmothers talked about. Nietzsche refers to this stage in his *Thus Spoke Zarathustra*: "There is always some madness in love." However, Nietzsche also added: "But there is also always some reason in madness." The latter might not be evident when we are truly, madly in love, but there is a reason for everything. The first stage of love is usually short, often lasting several months. During this stage, we get close to a stranger, and temporary madness likely facilitates this. When it wears off, we can proceed to the second stage, a passionate love.

After several months of being madly in love, the initial euphoria and excitement subdue. Stress is reduced and evolves into the feeling of safety, calmness, and fullness when we are with our partner. Passion and intimacy remain high, and commitment steadily increases. Oxytocin and vasopressin play vital roles during this stage, as love is slowly evolving into the attachment. Cortisol levels fall, and serotonin levels rise to normal, so we are no longer obsessed with love. Dopamine remains important, but feelings of being "high" are not as strong as they were.

Our brain adapts to the situation. We are becoming addicted to the closeness of our partners, but we no longer need excitement and euphoria. Calmness, trust, and a feeling of togetherness, caused by oxytocin and vasopressin, are more important now. They are also more rewarding than the excitement and euphoria triggered by dopamine. Our conscious brain kicks in again, and we are no longer blinded by love. We begin to think and function normally. But this brings along some problems. We begin to see our partners as they are and not as we have idolized them. Attachment grows, but so do differences that were covered up before. In this stage, it is vital that partners communicate, discuss all the differences, trust each other, and develop shared goals. Only then can the love progress into the third and the last stage, a companion love.

A companion love usually develops over several years. It grows on us. While the passion usually slowly fades, the intimacy and commitment remain high or even grow in a loving relationship. This stage of love is, in many respects, similar to a profound friendship. Partners are not mad about each other but rather become intimate friends. They do not go crazy when they are separated but simply know they can trust that the partner is waiting at home. Partners know each other well. They have gone through many things together, and they committed themselves to stay together for a very long time, perhaps for life. The fantasies from the initial madness are long gone. We know each other's faults, but we accept them. We argue occasionally, but we still love our partner. We sometimes miss the euphoria and excitement of being madly in love, and we might wonder if these feelings can be found again. But we are committed to our partner and are not seeking greener pastures. With the exceptions during occasional quarrels, our stress levels are low, and we feel calm and content. We sense we are in harmony with our partners.

Of course, every love story is different. Some people fall in love often, some only once in their lifetimes. For some, lust and mad love last only a few weeks, and after it subdues, they search for the same experience with new partners. Many partnerships break up during the second stage of love when differences between partners are not blinded any longer. Sometimes, the intimacy is low, and when passion placates, all that remains is a commitment, but this is not true love. It is sometimes called empty love. The duration of each stage is also different for every couple. Occasionally, we fall in love instantaneously, but we also forget this love in a short time. At other times, we might fall in love slowly. Sometimes we even fall in love with a person we have known for some time and have not initially imagined we could love them. Yet, sometimes, such slowly developing relationships develop into wonderful love stories.

There is one general rule: the first stage, being madly in love, never lasts. It might endure for several weeks or several months, but it will always subdue. When it does, we found ourselves on the crossroad: either break up the relationship or progress to passionate love. Passionate love lasts longer, usually several years, and in some couples, it never vanishes.

After passion has abated, couples either break up or progress to companionate love. It is often difficult to separate the passionate and companionate love. Many happy couples are primarily companions, but they still share the passion, even in their old age. Periods of more passionate and more companionate relationships are often swopping, probably together with the neurotransmitters' changes in our brains. When we are more passionate, oxytocin and vasopressin strengthen our relationships. We stay together during the periods of lower passion because intimacy and commitment remain high. Such true love can last for years if partners are willing to care for the relationship, just as we have to care for the garden. As every garden is different and unique because of its uniqueness, every relationship is special and unique. Each individual couple must find its own way how to preserve it. Yes, oxytocin and vasopressin trigger the feelings of attachment, intimacy, and commitment. But we have to nurture the love if we want it to blossom for a very long time.

Some divide the love into more than three stages, but gradual change in love is ever present. It is not the invention of western societies. !kung man, a member of a tribe of the San people living in the Kalahari Desert, once said to the anthropologist Marjorie Shostak while they observed a young !kung couple being madly in love: "When two people are first together, their hearts are on fire and their passion is very great. After a while, the fire cools and that's how it stays. They continue to love each other, but it's in a different way—warm and dependable." These words were uttered by a man with no connections to our society and no knowledge of our romantic literature. Yet the phrase could be easily recognized as typical for our western societies and our perception of love. This shows once again that love is a universal feeling, our faithful companion on our paths through life. The words of !kung man also show that love follows a similar pattern in every society, from the busy Times Square in New York to the remote and empty Kalahari Desert in Botswana.

Love has its biological origins in our distant evolutionary past. Bacteria started to exchange their genes, and this has led to the appearance of two sexes. When two sexes appeared, nature invented hormones and neurotransmitters to regulate the reproduction and pheromones for individuals of two sexes to find each other and reproduce. Once the vision has progressed in our ancestors, sight has replaced the pheromones, and for us, love goes partly through the eyes.

Sex and care for the offspring are present in many animals. As we climbed up on the evolutionary tree, both partnership and childcare were becoming more elaborate. Partnership and childcare started to be exceptionally important in birds and mammals. However, each group has followed its own independent path of evolutionary development. In mammals, gregariousness has become the prevalent mode of cohabitation. In birds, monogamy is present in around 90% of all species.

We have split apart from other apes during our progress from the common ancestors, and we followed different paths. Neither chimpanzees, bonobos, gorillas, nor orangutans show anything akin to the commitment and love between males and females, even though mothers take excellent care and love their children. Love has evolved in our species, most likely as babies became more dependent on maternal care, and mothers needed help to provide for their children. Rather than facing a risk of our genes being lost, men have evolved into romantic partners, taking care of their lovers and children.

We can find parallels between human love and attachment and pair bonding in animals. Prairie voles mate for life, as do gibbons and swans. Penguins remain faithful to each other even when separated, and marmosets share their lives till death do them part. In all mammals, oxytocin and vasopressin, or their equivalents in birds, are responsible for attachment and pair bonding. But human love is certainly not only hormones and neurotransmitters. We are not just flesh and blood. We are not only instincts. Besides the reptilian brain, we have our unique, conscious mind. While reptiles mate and prairie voles stay together, we are the only species singing love songs and writing romantic novels. We are writing sonnets and pretend to be the troubadours when we try to win our beloved's heart.

We do not know what love is, and we might never understand. Our brain is the most complicated structure in the known universe. Although we are slowly learning how to glimpse into the functioning of our brain, we are far from understanding how the mind works. This is especially true for things like love. We know that we have more dopamine in our brain when we are in love. We understand that there is more oxytocin when we feel close to our partner. We comprehend that vasopressin is released when we are sexually aroused. We learned that serotonin levels fall when we are madly in love. But how all these brain activities produce the feeling of love and attachment is beyond our comprehension, and it will likely remain so for the foreseeable future.

The perpetual cycle of life is one of the most wonderful mysteries of our planet. We eat, grow, work, and think. But most importantly, we reproduce, and this sustains us and our kin alive. The essence of life, our genes, never die. They renew in each generation. Our cells grow old, our skin wrinkles, our hearts weaken, our hair thins, and our muscles shrink. But our genes are constantly renewed and repaired.

Combinations of genes change through the unique genetic blending in each generation, but essentially, genes remain the same. We carry the same genes as Adam and Eve did in the paradise, similar genes were present in the Lucy who roamed the Ethiopian plains more than 3 million years ago. We do not need to strive for eternal life, because we already have it! Through love, one of the most powerful emotions, we can achieve immortality.

Therefore, we should not seek eternal life for ourselves but rather strive to fill the one we have with love and compassion. Every child raised in a lovable relationship will grow up to be a lovable person who will love and project love. With each such generation, there will be less violence and sadness in this world, and perhaps one day, we will live in the world without killing and gruesomeness. We are building such a world not only for our children and grandchildren. We are making it for ourselves, for us, who are living in our children and grandchildren.

Love is a flow of hormones through our veins and sparkling nerve cells in our brains. But it is not only that. It is the essence of life. It has evolved to give us eternal life. Through love for our partners and love for our children, our mission on earth is accomplished. Genes and habits, roots and wings, given to our children through love, will defy the decay of time. So even though love first evolved as nature's trick to force us to procreate, love is not only a cocktail of hormones and neurotransmitters. We have learned a lot about what is happening inside our brains when we love. Yet, it is still unfathomable for us to understand the most complicated organ in the known universe, our brain. It is becoming more and more apparent that the brain is more than just sparkling of billions of nerve cells. That is why we will probably never completely decipher the recipe for love, and that is just as well. Yes, scientists should explore what is going on in our brains when we love. But part of the love is and should remain beyond the reach of science, as it is in the realm of bards.

# Further Reading

Abou Chakra M, Hilbe C, Traulsen A (2016) Coevolutionary interactions between farmers and mafia induce host acceptance of avian brood parasites. R Soc Open Sci 3 (5):160036. https://doi.org/10.1098/rsos.160036

Abraham DS, Gruber-Baldini AL, Magder LS, McArdle PF, Tom SE, Barr E, Schrader K, Shulman LM (2019) Sex differences in Parkinson's disease presentation and progression. Parkinsonism Relat Disord 69:48–54. https://doi.org/10.1016/j.parkreldis.2019.10.019

Adler J (1978) Chemotaxis in bacteria. Harvey Lect 72:195–230

Albers HE (2012) The regulation of social recognition, social communication and aggression: vasopressin in the social behavior neural network. Horm Behav 61 (3):283–292. https://doi.org/10.1016/j.yhbeh.2011.10.007

Albert DJ, Walsh ML, Jonik RH (1993) Aggression in humans: what is its biological foundation? Neurosci Biobehav Rev 17 (4):405–425. https://doi.org/10.1016/s0149-7634(05)80117-4

Allen JJ, Anderson CA (2017) Aggression and Violence: Definitions and Distinctions. In: Sturmey P (ed) The Wiley Handbook of Violence and Aggression, vol 1. Wiley-Blackwell, Hoboken, NJ, pp. 1–14

Allott R (1992) Evolutionary aspects of love and empathy. Journal of Social and Evolutionary Systems 15 (4):353–370. https://doi.org/10.1016/1061-7361(92)90023-7

Amcoff M, Gonzalez-Voyer A, Kolm N (2013) Evolution of egg dummies in Tanganyikan cichlid fishes: the roles of parental care and sexual selection. J Evol Biol 26 (11):2369–2382. https://doi.org/10.1111/jeb.12231

Anacker AM, Beery AK (2013) Life in groups: the roles of oxytocin in mammalian sociality. Front Behav Neurosci 7:185. https://doi.org/10.3389/fnbeh.2013.00185

Arnold AP, Rissman EF, De Vries GJ (2003) Two perspectives on the origin of sex differences in the brain. Ann N Y Acad Sci 1007:176–188. https://doi.org/10.1196/annals.1286.018

Bachtrog D, Mank JE, Peichel CL, Kirkpatrick M, Otto SP, Ashman T-L, Hahn MW, Kitano J, Mayrose I, Ming R, Perrin N, Ross L, Valenzuela N, Vamosi JC, The Tree of Sex C (2014) Sex Determination: Why So Many Ways of Doing It? PLOS Biology 12 (7):e1001899. https://doi.org/10.1371/journal.pbio.1001899

Bailey NW, Zuk M (2009) Same-sex sexual behavior and evolution. Trends Ecol Evol 24 (8):439–446. https://doi.org/10.1016/j.tree.2009.03.014

Ball GF, Riters LV, Balthazart J (2002) Neuroendocrinology of song behavior and avian brain plasticity: multiple sites of action of sex steroid hormones. Front Neuroendocrinol 23 (2):137–178. https://doi.org/10.1006/frne.2002.0230

Balthazart J (2016) Sex differences in partner preferences in humans and animals. Philos Trans R Soc Lond B Biol Sci 371 (1688):20150118. https://doi.org/10.1098/rstb.2015.0118



G. Majdic, *Soul Mate Biology*, Fascinating Life Sciences,
https://doi.org/10.1007/978-3-030-67212-6

Balthazart J (2020) Sexual partner preference in animals and humans. Neurosci Biobehav Rev 115:34–47. https://doi.org/10.1016/j.neubiorev.2020.03.024

Barras C (2014) Only known chimp war reveals how societies splinter. New scientist 222 (2968):12

Baylis JR (1981) The evolution of parental care in fishes, with reference to Darwin's rule of male sexual selection. Environmental Biology of Fishes 6 (2):223–251. https://doi.org/10.1007/BF00002788

Beauregard M, Courtemanche J, Paquette V, St-Pierre EL (2009) The neural basis of unconditional love. Psychiatry Res 172 (2):93–98. https://doi.org/10.1016/j.pscychresns.2008.11.003

Becker JB, Arnold AP, Berkley KJ, Blaustein JD, Eckel LA, Hampson E, Herman JP, Marts S, Sadee W, Steiner M, Taylor J, Young E (2005) Strategies and methods for research on sex differences in brain and behavior. Endocrinology 146 (4):1650–1673. https://doi.org/10.1210/en.2004-1142

Ben-Ami Bartal I, Rodgers DA, Bernardez Sarria MS, Decety J, Mason P (2014) Pro-social behavior in rats is modulated by social experience. eLife 3:e01385. https://doi.org/10.7554/eLife.01385

Benun Sutton F, Wilson AB (2019) Where are all the moms? External fertilization predicts the rise of male parental care in bony fishes. Evolution 73 (12):2451–2460. https://doi.org/10.1111/evo.13846

Bilde T, Tuni C, Elsayed R, Pekar S, Toft S (2006) Death feigning in the face of sexual cannibalism. Biol Lett 2 (1):23–25. https://doi.org/10.1098/rsbl.2005.0392

Bingman VP, Gagliardo A (2006) Of birds and men: convergent evolution in hippocampal lateralization and spatial cognition. Cortex 42 (1):99–100. https://doi.org/10.1016/s0010-9452(08)70329-0

Birkhead TR, Mølle AP (1998) Sperm Competition and Sexual Selection. Academic Press, London,

Bjornerem A, Straume B, Oian P, Berntsen GK (2006) Seasonal variation of estradiol, follicle stimulating hormone, and dehydroepiandrosterone sulfate in women and men. J Clin Endocrinol Metab 91 (10):3798–3802. https://doi.org/10.1210/jc.2006-0866

Blaustein J (2007) Behavioral neurochemistry, neuroendocrinology and molecular neurobiology. Handbook of neurochemistry and molecular neurobiology (3rd ed). Springer, New York; Great Britain

Bonnet L, Comte A, Tatu L, Millot JL, Moulin T, Medeiros de Bustos E (2015) The role of the amygdala in the perception of positive emotions: an "intensity detector". Front Behav Neurosci 9:178. https://doi.org/10.3389/fnbeh.2015.00178

Bredewold R, Veenema AH (2018) Sex differences in the regulation of social and anxiety-related behaviors: insights from vasopressin and oxytocin brain systems. Curr Opin Neurobiol 49:132–140. https://doi.org/10.1016/j.conb.2018.02.011

Brennan PA (2010) Pheromones and Mammalian Behavior. In: Menini A (ed) The Neurobiology of Olfaction. CRC Press/Taylor & Francis, Boca Raton (FL),

Briffa M (2010) Territoriality and Aggression. Nature Education Knowledge 3 (10):81

Buss DM (2018) The Evolution of Love in Humans. In: Sternberg K, Sternberg RJ (eds) The New Psychology of Love. 2 edn. Cambridge University Press, Cambridge, pp. 42-63. https://doi.org/10.1017/9781108658225.004

Cacioppo S, Bianchi-Demicheli F, Frum C, Pfaus JG, Lewis JW (2012a) The Common Neural Bases Between Sexual Desire and Love: A Multilevel Kernel Density fMRI Analysis. The Journal of Sexual Medicine 9 (4):1048–1054. https://doi.org/10.1111/j.1743-6109.2012.02651.x

Cacioppo S, Bianchi-Demicheli F, Frum C, Pfaus JG, Lewis JW (2012b) The Common Neural Bases Between Sexual Desire and Love: A Multilevel Kernel Density fMRI Analysis. The Journal of Sexual Medicine 9 (4):1048–1054. https://doi.org/10.1111/j.1743-6109.2012.02651.x

Caldwell RL, Ross R, Rodaniche A, Huffard CL (2015) Behavior and Body Patterns of the Larger Pacific Striped Octopus. PLoS One 10 (8):e0134152. https://doi.org/10.1371/journal.pone.0134152

Carr B (2018) Attention women: Low sex drive? Testosterone won't help. https://utswmed.org/medblog/libido-testosterone-women/. 2020

Carre JM, Archer J (2018) Testosterone and human behavior: the role of individual and contextual variables. Curr Opin Psychol 19:149–153. https://doi.org/10.1016/j.copsyc.2017.03.021

Carter CS (2017a) The Oxytocin-Vasopressin Pathway in the Context of Love and Fear. Front Endocrinol (Lausanne) 8:356. https://doi.org/10.3389/fendo.2017.00356

Carter CS (2017b) The Role of Oxytocin and Vasopressin in Attachment. Psychodyn Psychiatry 45 (4):499–517. https://doi.org/10.1521/pdps.2017.45.4.499

Choleris E, Galea LAM, Sohrabji F, Frick KM (2018) Sex differences in the brain: Implications for behavioral and biomedical research. Neurosci Biobehav Rev 85:126–145. https://doi.org/10.1016/j.neubiorev.2017.07.005

Curlik DM, 2nd, Difeo G, Shors TJ (2014) Preparing for adulthood: thousands upon thousands of new cells are born in the hippocampus during puberty, and most survive with effortful learning. Front Neurosci 8:70. https://doi.org/10.3389/fnins.2014.00070

de Vries T, Lakes-Harlan R (2007) Prenatal cannibalism in an insect. Naturwissenschaften 94 (6):477–482. https://doi.org/10.1007/s00114-006-0213-z

Della Torre S, Maggi A (2017) Sex Differences: A Resultant of an Evolutionary Pressure? Cell Metab 25 (3):499–505. https://doi.org/10.1016/j.cmet.2017.01.006

DiBenedictis BT, Cheung HK, Nussbaum ER, Veenema AH (2020) Involvement of ventral pallidal vasopressin in the sex-specific regulation of sociosexual motivation in rats. Psychoneuroendocrinology 111:104462. https://doi.org/10.1016/j.psyneuen.2019.104462

Disorders IoMUFoNaNS (2011) Sex Differences and Implications for Translational Neuroscience Research: Workshop Summary. National Academies Press, Washington (DC). https://doi.org/10.17226/13004

Dixson AL, Anderson MJ (2002) Sexual selection, seminal coagulation and copulatory plug formation in primates. Folia Primatol (Basel) 73 (2–3):63–69. https://doi.org/10.1159/000064784

Doving KB, Trotier D (1998) Structure and function of the vomeronasal organ. J Exp Biol 201 (Pt 21):2913–2925

Duchemin A, Seelke AM, Simmons TC, Freeman SM, Bales KL (2017) Localization of oxytocin receptors in the prairie vole (Microtus ochrogaster) neocortex. Neuroscience 348:201–211. https://doi.org/10.1016/j.neuroscience.2017.02.017

Dumais KM, Alonso AG, Bredewold R, Veenema AH (2016) Role of the oxytocin system in amygdala subregions in the regulation of social interest in male and female rats. Neuroscience 330:138–149. https://doi.org/10.1016/j.neuroscience.2016.05.036

Eastwick PW (2009) Beyond the pleistocene: Using phylogeny and constraint to inform the evolutionary psychology of human mating. Psychological Bulletin 135 (5):794–821

Ebstein RP, Israel S, Lerer E, Uzefovsky F, Shalev I, Gritsenko I, Riebold M, Salomon S, Yirmiya N (2009) Arginine vasopressin and oxytocin modulate human social behavior. Ann N Y Acad Sci 1167:87–102. https://doi.org/10.1111/j.1749-6632.2009.04541.x

Edmunds M (1988) Sexual cannibalism in mantids. Trends Ecol Evol 3 (3):77. https://doi.org/10.1016/0169-5347(88)90021-3

Eichelman B (1983) The limbic system and aggression in humans. Neurosci Biobehav Rev 7 (3):391–394. https://doi.org/10.1016/0149-7634(83)90044-1

Estes RD (2012) The behavior guide to african mammals. University of California Press, Los Angeles, California

Fallon BL (2016) Courtship Communication in the Wild Chimpanzees of Budongo, Uganda. University of St Andrews, St Andrews

Farrell AP, Stevens ED, Cech JJ, Richards JG (2011) Encyclopedia of fish physiology: from genome to environment. Academic Press, an imprint of Elsevier, London; Waltham, MA

Feldman R, Monakhov M, Pratt M, Ebstein RP (2016) Oxytocin Pathway Genes: Evolutionary Ancient System Impacting on Human Affiliation, Sociality, and Psychopathology. Biol Psychiatry 79 (3):174–184. https://doi.org/10.1016/j.biopsych.2015.08.008

Filippi S, Vignozzi L, Vannelli GB, Ledda F, Forti G, Maggi M (2003) Role of oxytocin in the ejaculatory process. J Endocrinol Invest 26 (3 Suppl):82–86

Fisher HE, Xu X, Aron A, Brown LL (2016) Intense, Passionate, Romantic Love: A Natural Addiction? How the Fields That Investigate Romance and Substance Abuse Can Inform Each Other. Front Psychol 7:687. https://doi.org/10.3389/fpsyg.2016.00687

Floody OR, Pfaff DW (1974) Steroid hormones and aggressive behavior: approaches to the study of hormone-sensitive brain mechanisms for behavior. Res Publ Assoc Res Nerv Ment Dis 52:149–185

Forbes EE, Dahl RE (2010) Pubertal development and behavior: hormonal activation of social and motivational tendencies. Brain Cogn 72 (1):66–72. https://doi.org/10.1016/j.bandc.2009.10.007

Frankl-Vilches C, Gahr M (2018) Androgen and estrogen sensitivity of bird song: a comparative view on gene regulatory levels. J Comp Physiol A Neuroethol Sens Neural Behav Physiol 204 (1):113–126. https://doi.org/10.1007/s00359-017-1236-y

Ganswindt A, Rasmussen HB, Heistermann M, Hodges JK (2005) The sexually active states of free-ranging male African elephants (Loxodonta africana): defining musth and non-musth using endocrinology, physical signals, and behavior. Horm Behav 47 (1):83–91. https://doi.org/10.1016/j.yhbeh.2004.09.002

Geniole SN, Bird BM, McVittie JS, Purcell RB, Archer J, Carré JM (2019) Is testosterone linked to human aggression? A meta-analytic examination of the relationship between baseline, dynamic, and manipulated testosterone on human aggression. Hormones and Behavior:104644. https://doi.org/10.1016/j.yhbeh.2019.104644

Gilks WP, Abbott JK, Morrow EH (2014) Sex differences in disease genetics: evidence, evolution, and detection. Trends Genet 30 (10):453–463. https://doi.org/10.1016/j.tig.2014.08.006

Gillies GE, Pienaar IS, Vohra S, Qamhawi Z (2014) Sex differences in Parkinson's disease. Front Neuroendocrinol 35 (3):370–384. https://doi.org/10.1016/j.yfrne.2014.02.002

Giuggioli L, Potts JR, Harris S (2011) Animal interactions and the emergence of territoriality. PLoS Comput Biol 7 (3):e1002008. https://doi.org/10.1371/journal.pcbi.1002008

Godwin J, Thompson R (2012) Nonapeptides and social behavior in fishes. Horm Behav 61 (3):230–238. https://doi.org/10.1016/j.yhbeh.2011.12.016

Gonzalez-Voyer A, Fitzpatrick JL, Kolm N (2008) Sexual selection determines parental care patterns in cichlid fishes. Evolution 62 (8):2015–2026. https://doi.org/10.1111/j.1558-5646.2008.00426.x

Goodson JL, Kelly AM, Kingsbury MA (2012) Evolving nonapeptide mechanisms of gregariousness and social diversity in birds. Horm Behav 61 (3):239–250. https://doi.org/10.1016/j.yhbeh.2012.01.005

Gottfried JA (2011) Neurobiology of Sensation and Reward. CRC Press/Taylor & Francis, Boca Raton (FL)

Haller J (2017) Studies into abnormal aggression in humans and rodents: Methodological and translational aspects. Neurosci Biobehav Rev. 76 (Pt A):77-86. https://doi.org/10.1016/j.neubiorev.2017.02.022

Haller J, Harold G, Sandi C, Neumann ID (2014) Effects of adverse early-life events on aggression and anti-social behaviours in animals and humans. J Neuroendocrinol 26 (10):724–738. https://doi.org/10.1111/jne.12182

Harrison D (1987) Preliminary thoughts on the incidence, structure and function of the mammalian vomeronasal organ. Acta Otolaryngol 103 (5–6):489–495

Hartgens F, Kuipers H (2004) Effects of androgenic-anabolic steroids in athletes. Sports Med 34 (8):513–554. https://doi.org/10.2165/00007256-200434080-00003

Hazelbauer GL (2012) Bacterial chemotaxis: the early years of molecular studies. Annu Rev. Microbiol 66:285–303. https://doi.org/10.1146/annurev-micro-092611-150120

Heinrichs M, von Dawans B, Domes G (2009) Oxytocin, vasopressin, and human social behavior. Front Neuroendocrinol 30 (4):548–557. https://doi.org/10.1016/j.yfrne.2009.05.005

Herbison AE (2016) Control of puberty onset and fertility by gonadotropin-releasing hormone neurons. Nat Rev. Endocrinol 12 (8):452–466. https://doi.org/10.1038/nrendo.2016.70

Hospenthal MK, Costa TRD, Waksman G (2017) A comprehensive guide to pilus biogenesis in Gram-negative bacteria. Nat Rev. Microbiol 15 (6):365–379. https://doi.org/10.1038/nrmicro.2017.40

Hueston CM, Cryan JF, Nolan YM (2017) Stress and adolescent hippocampal neurogenesis: diet and exercise as cognitive modulators. Transl Psychiatry 7 (4):e1081. https://doi.org/10.1038/tp.2017.48

Hussain YH, Guasto JS, Zimmer RK, Stocker R, Riffell JA (2016) Sperm chemotaxis promotes individual fertilization success in sea urchins. J Exp Biol 219 (Pt 10):1458–1466. https://doi.org/10.1242/jeb.134924

Insall R, Andrew N (2007) Chemotaxis in Dictyostelium: how to walk straight using parallel pathways. Curr Opin Microbiol 10 (6):578–581. https://doi.org/10.1016/j.mib.2007.10.004

Jacobs LF, Gaulin SJ, Sherry DF, Hoffman GE (1990) Evolution of spatial cognition: sex-specific patterns of spatial behavior predict hippocampal size. Proc Natl Acad Sci U S A 87 (16):6349–6352. https://doi.org/10.1073/pnas.87.16.6349

Jansson-Knodell CL, Hujoel IA, West CP, Taneja V, Prokop LJ, Rubio-Tapia A, Murray JA (2019) Sex Difference in Celiac Disease in Undiagnosed Populations: A Systematic Review and Meta-analysis. Clin Gastroenterol Hepatol 17 (10):1954–1968.e13. https://doi.org/10.1016/j.cgh.2018.11.013

Johns P (2014) Chapter 3 - Functional neuroanatomy. In: Johns P (ed) Clinical Neuroscience. Churchill Livingstone, pp. 27-47. https://doi.org/10.1016/B978-0-443-10321-6.00003-5

Johnson JC, Ivy TM, Sakaluk SK (1999) Female remating propensity contingent on sexual cannibalism in sagebrush crickets, Cyphoderris strepitans: a mechanism of cryptic female choice. Behavioral Ecology 10 (3):227–233. https://doi.org/10.1093/beheco/10.3.227

Karlson P, Luscher M (1959) Pheromones': a new term for a class of biologically active substances. Nature 183 (4653):55–56. https://doi.org/10.1038/183055a0

Kawada A, Nagasawa M, Murata A, Mogi K, Watanabe K, Kikusui T, Kameda T (2019) Vasopressin enhances human preemptive strike in both males and females. Scientific Reports 9 (1):9664. https://doi.org/10.1038/s41598-019-45953-y

Kelly DR (1996) When is a butterfly like an elephant? Chemistry & Biology 3 (8):595–602. https://doi.org/10.1016/S1074-5521(96)90125-8

Kim P, Strathearn L, Swain JE (2016) The maternal brain and its plasticity in humans. Horm Behav 77:113–123. https://doi.org/10.1016/j.yhbeh.2015.08.001

Kim S, Strathearn L (2016) Oxytocin and Maternal Brain Plasticity. New Dir Child Adolesc Dev 2016 (153):59–72. https://doi.org/10.1002/cad.20170

Komers PE (1996) Obligate monogamy without paternal care in Kirk's dikdik. Animal Behaviour 51 (1):131–140. https://doi.org/10.1006/anbe.1996.0011.

Kompier NF, Keysers C, Gazzola V, Lucassen PJ, Krugers HJ (2019) Early Life Adversity and Adult Social Behavior: Focus on Arginine Vasopressin and Oxytocin as Potential Mediators. Front Behav Neurosci 13:143. https://doi.org/10.3389/fnbeh.2019.00143

Kutlu MG, Burke D, Slade S, Hall BJ, Rose JE, Levin ED (2013) Role of insular cortex D(1) and D(2) dopamine receptors in nicotine self-administration in rats. Behav Brain Res 256:273–278. https://doi.org/10.1016/j.bbr.2013.08.005

Kvarnemo C (2018) Why do some animals mate with one partner rather than many? A review of causes and consequences of monogamy. Biological Reviews (93):1795–1812. https://doi.org/10.1111/brv.12421

Lam CSP, Arnott C, Beale AL, Chandramouli C, Hilfiker-Kleiner D, Kaye DM, Ky B, Santema BT, Sliwa K, Voors AA (2019) Sex differences in heart failure. Eur Heart J 40 (47):3859-3868c. https://doi.org/10.1093/eurheartj/ehz835

Lavin C, Melis C, Mikulan E, Gelormini C, Huepe D, Ibanez A (2013) The anterior cingulate cortex: an integrative hub for human socially-driven interactions. Front Neurosci 7:64. https://doi.org/10.3389/fnins.2013.00064

Lawrence JG (2002) Gene Transfer in Bacteria: Speciation without Species? Theoretical Population Biology 61 (4):449–460. https://doi.org/10.1006/tpbi.2002.1587

Laws KR, Irvine K, Gale TM (2018) Sex differences in Alzheimer's disease. Curr Opin Psychiatry 31 (2):133–139. https://doi.org/10.1097/YCO.0000000000000401

Lee TT, Wainwright SR, Hill MN, Galea LA, Gorzalka BB (2014) Sex, drugs, and adult neurogenesis: sex-dependent effects of escalating adolescent cannabinoid exposure on adult hippocampal neurogenesis, stress reactivity, and amphetamine sensitization. Hippocampus 24 (3):280–292. https://doi.org/10.1002/hipo.22221

Lenroot RK, Giedd JN (2010) Sex differences in the adolescent brain. Brain Cogn 72 (1):46–55. https://doi.org/10.1016/j.bandc.2009.10.008

Library WPRC (2020) Primate Factsheets. http://pin.primate.wisc.edu/factsheets/. 2020

Lorenz MG, Wackernagel W (1994) Bacterial gene transfer by natural genetic transformation in the environment. Microbiol Rev 58 (3):563–602

MacLean PD (1977) The triune brain in conflict. Psychother Psychosom 28 (1–4):207–220. https://doi.org/10.1159/000287065

Marazziti D, Canale D (2004) Hormonal changes when falling in love. Psychoneuroendocrinology 29 (7):931–936. https://doi.org/10.1016/j.psyneuen.2003.08.006

McCarthy MM, Arnold AP, Ball GF, Blaustein JD, De Vries GJ (2012) Sex differences in the brain: the not so inconvenient truth. J Neurosci 32 (7):2241–2247. https://doi.org/10.1523/JNEUROSCI.5372-11.2012

McClintock MK (1971) Menstrual Synchrony and Suppression. Nature 229 (5282):244–245. https://doi.org/10.1038/229244a0

McEwen BS (2020) Hormones and behavior and the integration of brain-body science. Horm Behav 119:104619. https://doi.org/10.1016/j.yhbeh.2019.104619

Meyza KZ, Bartal IB-A, Monfils MH, Panksepp JB, Knapska E (2017) The roots of empathy: Through the lens of rodent models. Neuroscience & Biobehavioral Reviews 76:216–234. https://doi.org/10.1016/j.neubiorev.2016.10.028

Michael RP, Zumpe D (1983) Sexual violence in the United States and the role of season. Am J Psychiatry 140 (7):883–886. https://doi.org/10.1176/ajp.140.7.883

Mielke MM (2018) Sex and Gender Differences in Alzheimer's Disease Dementia. Psychiatr Times 35 (11):14–17

Miller BS, Kendall AW (2009) Early life history of marine fishes. University of California Press, Berkeley, Calif.

Miller WL (2018) The Hypothalamic-Pituitary-Adrenal Axis: A Brief History. Horm Res Paediatr 89 (4):212–223. https://doi.org/10.1159/000487755

Monk JD, Giglio E, Kamath A, Lambert MR, McDonough CE (2019) An alternative hypothesis for the evolution of same-sex sexual behaviour in animals. Nat Ecol Evol 3 (12):1622–1631. https://doi.org/10.1038/s41559-019-1019-7

Morrow EH (2015) The evolution of sex differences in disease. Biol Sex Differ 6:5. https://doi.org/10.1186/s13293-015-0023-0

Morse P, Huffard CL (2019) Tactical Tentacles: New Insights on the Processes of Sexual Selection Among the Cephalopoda. Front Physiol 10:1035. https://doi.org/10.3389/fphys.2019.01035

Najman JM, Hayatbakhsh MR, McGee TR, Bor W, O'Callaghan MJ, Williams GM (2009) The Impact of Puberty on Aggression/Delinquency: Adolescence to Young Adulthood. Australian & New Zealand Journal of Criminology 42 (3):369–386. https://doi.org/10.1375/acri.42.3.369

Natarajan D, Caramaschi D (2010) Animal violence demystified. Front Behav Neurosci 4:9. https://doi.org/10.3389/fnbeh.2010.00009

Naylor R, Richardson SJ, McAllan BM (2008) Boom and bust: a review of the physiology of the marsupial genus Antechinus. J Comp Physiol B 178 (5):545–562. https://doi.org/10.1007/s00360-007-0250-8

Nelson RJ, Kriegsfeld LJ (2017) An introduction to behavioral endocrinology. Fifth edition. Edn. Sinauer Associates, Inc. Publishers, Sunderland, Massachusetts

Newman SW (1999) The medial extended amygdala in male reproductive behavior. A node in the mammalian social behavior network. Ann N Y Acad Sci 877:242–257. https://doi.org/10.1111/j.1749-6632.1999.tb09271.x

Nieschlag E, Nieschlag S (2019) ENDOCRINE HISTORY: The history of discovery, synthesis and development of testosterone for clinical use. Eur J Endocrinol 180 (6):R201-R212. https://doi.org/10.1530/EJE-19-0071

Nieschlag E, Vorona E (2015) Doping with anabolic androgenic steroids (AAS): Adverse effects on non-reproductive organs and functions. Rev Endocr Metab Disord 16 (3):199–211. https://doi.org/10.1007/s11154-015-9320-5

Numan M, Insel TR (2003) The neurobiology of parental behavior. Hormones, brain, and behavior. Springer, New York

Ogunseitan OA (1995) Bacterial genetic exchange in nature. Sci Prog 78 (Pt 3):183–204

Opie C, Atkinson QD, Dunbar RI, Shultz S (2013) Male infanticide leads to social monogamy in primates. Proc Natl Acad Sci U S A 110 (33):13328–13332. https://doi.org/10.1073/pnas.1307903110

Oxley G, Fleming AS (2000) The effects of medial preoptic area and amygdala lesions on maternal behavior in the juvenile rat. Dev Psychobiol 37 (4):253–265

Partridge SR, Kwong SM, Firth N, Jensen SO (2018) Mobile Genetic Elements Associated with Antimicrobial Resistance. Clin Microbiol Rev 31 (4). https://doi.org/10.1128/CMR.00088-17

Pascini TV, Martins GF (2017) The insect spermatheca: an overview. Zoology (Jena) 121:56–71. https://doi.org/10.1016/j.zool.2016.12.001

Pavlov KA, Chistiakov DA, Chekhonin VP (2012) Genetic determinants of aggression and impulsivity in humans. J Appl Genet 53 (1):61–82. https://doi.org/10.1007/s13353-011-0069-6

Pedersen CA, Chang SW, Williams CL (2014) Evolutionary perspectives on the role of oxytocin in human social behavior, social cognition and psychopathology. Brain Res 1580:1–7. https://doi.org/10.1016/j.brainres.2014.07.033

Peper JS, Dahl RE (2013) Surging Hormones: Brain-Behavior Interactions During Puberty. Curr Dir Psychol Sci 22 (2):134–139. https://doi.org/10.1177/0963721412473755

Perrone M, Zaret TM (1979) Parental Care Patterns of Fishes. The American Naturalist 113 (3):351–361

Pessoa L, Hof PR (2015) From Paul Broca's great limbic lobe to the limbic system. J Comp Neurol 523 (17):2495–2500. https://doi.org/10.1002/cne.23840

Pfaff DW, Joëls M (2017) Hormones, brain, and behavior. Third edition. Edn. Elsevier/AP, Academic Press is an imprint of Elsevier, Amsterdam; Boston;

Pfaff DW, Volkow ND (2016) Neuroscience in the twenty-first century: from basic to clinical. Springer reference, Second edition. Edn. Springer, New York

Pittman KJ, Glover LC, Wang L, Ko DC (2016) The Legacy of Past Pandemics: Common Human Mutations That Protect against Infectious Disease. PLoS Pathog 12 (7):e1005680. https://doi.org/10.1371/journal.ppat.1005680

Price J, Allen S (2004) Exploring the mechanisms regulating regeneration of deer antlers. Philos Trans R Soc Lond B Biol Sci 359 (1445):809–822. https://doi.org/10.1098/rstb.2004.1471

Regitz-Zagrosek V, Seeland U (2011) Sex and gender differences in myocardial hypertrophy and heart failure. Wien Med Wochenschr 161 (5–6):109–116. https://doi.org/10.1007/s10354-011-0892-8

Reichard UH, Boesch C (2003) Monogamy: Mating Strategies and Partnerships in Birds, Humans and Other Mammals. Cambridge University Press,

Reiner A (1990) The Triune Brain in Evolution. Role in Paleocerebral Functions. Paul D. MacLean. Plenum, New York, 1990. xxiv, 672 pp., illus. $75. Science 250 (4978):303–305. https://doi.org/10.1126/science.250.4978.303-a

Roberts EK, Lu A, Bergman TJ, Beehner JC (2012) A Bruce effect in wild geladas. Science 335 (6073):1222–1225. https://doi.org/10.1126/science.1213600

Rock D, Greenberg DM, Hallmayer J (2003) Cyclical changes of homicide rates: a reanalysis of Brearley's 1932 data. J Interpers Violence 18 (8):942–955. https://doi.org/10.1177/0886260503254404

Roselli CE, Stormshak F (2009) Prenatal programming of sexual partner preference: the ram model. J Neuroendocrinol 21 (4):359–364. https://doi.org/10.1111/j.1365-2826.2009.01828.x

Ruesink GB, Georgiadis JR (2017) Brain Imaging of Human Sexual Response: Recent Developments and Future Directions. Curr Sex Health Rep 9 (4):183–191. https://doi.org/10.1007/s11930-017-0123-4

Santen RJ, Simpson E (2019) History of Estrogen: Its Purification, Structure, Synthesis, Biologic Actions, and Clinical Implications. Endocrinology 160 (3):605–625. https://doi.org/10.1210/en.2018-00529

Schacht R, Bell AV (2016) The evolution of monogamy in response to partner scarcity. Sci Rep 6:32472. https://doi.org/10.1038/srep32472

Schacht R, Kramer KL (2019) Are We Monogamous? A Review of the Evolution of Pair-Bonding in Humans and Its Contemporary Variation Cross-Culturally. Frontiers in Ecology and Evolution 7 (230). https://doi.org/10.3389/fevo.2019.00230

Schneider JM (2014) Sexual cannibalism as a manifestation of sexual conflict. Cold Spring Harb Perspect Biol 6 (11):a017731. https://doi.org/10.1101/cshperspect.a017731

Schwartz SK, Wagner WE, Jr., Hebets EA (2016) Males Can Benefit from Sexual Cannibalism Facilitated by Self-Sacrifice. Curr Biol 26 (20):2794–2799. https://doi.org/10.1016/j.cub.2016.08.010

Shaw K (2012) How Aggressive Hyena Moms Give Their Kids a Boost. Scientific American

Sheehan T, Paul M, Amaral E, Numan MJ, Numan M (2001) Evidence that the medial amygdala projects to the anterior/ventromedial hypothalamic nuclei to inhibit maternal behavior in rats. Neuroscience 106 (2):341–356. https://doi.org/10.1016/S0306-4522(01)00286-X

Simonyan K, Horwitz B, Jarvis ED (2012) Dopamine regulation of human speech and bird song: a critical review. Brain Lang 122 (3):142–150. https://doi.org/10.1016/j.bandl.2011.12.009

Sisk CL, Zehr JL (2005) Pubertal hormones organize the adolescent brain and behavior. Front Neuroendocrinol 26 (3–4):163–174. https://doi.org/10.1016/j.yfrne.2005.10.003

Smirnova T, Segall JE (2007) Amoeboid chemotaxis: future challenges and opportunities. Cell Adh Migr 1 (4):165–170. https://doi.org/10.4161/cam.1.4.5305

Smith CU (2010) The triune brain in antiquity: Plato, Aristotle, Erasistratus. J Hist Neurosci 19 (1):1–14. https://doi.org/10.1080/09647040802601605

Smith KS, Tindell AJ, Aldridge JW, Berridge KC (2009) Ventral pallidum roles in reward and motivation. Behav Brain Res 196 (2):155–167. https://doi.org/10.1016/j.bbr.2008.09.038

Sorokowski P, Zelazniewicz A, Nowak J, Groyecka A, Kaleta M, Lech W, Samorek S, Stachowska K, Bocian K, Pulcer A, Sorokowska A, Kowal M, Pisanski K (2019) Romantic Love and Reproductive Hormones in Women. Int J Environ Res Public Health 16 (21). https://doi.org/10.3390/ijerph16214224

Stanton SJ, Mullette-Gillman OA, Huettel SA (2011) Seasonal variation of salivary testosterone in men, normally cycling women, and women using hormonal contraceptives. Physiol Behav 104 (5):804–808. https://doi.org/10.1016/j.physbeh.2011.07.009

Stern K, McClintock MK (1998) Regulation of ovulation by human pheromones. Nature 392 (6672):177–179. https://doi.org/10.1038/32408

Sternberg RJ, Sternberg K (2019) The new psychology of love. Second edition. Edn. Cambridge University Press, Cambridge, United Kingdom; New York, NY

Stevens FL, Hurley RA, Taber KH (2011) Anterior cingulate cortex: unique role in cognition and emotion. J Neuropsychiatry Clin Neurosci 23 (2):121-125. https://doi.org/10.1176/appi.neuropsych.23.2.121. https://doi.org/10.1176/jnp.23.2.jnp121

Strassmann BI (1999) Menstrual synchrony pheromones: cause for doubt. Hum Reprod 14 (3):579–580. https://doi.org/10.1093/humrep/14.3.579

Szymanska M, Schneider M, Chateau-Smith C, Nezelof S, Vulliez-Coady L (2017) Psychophysiological effects of oxytocin on parent-child interactions: A literature review on oxytocin and parent-child interactions. Psychiatry Clin Neurosci 71 (10):690–705. https://doi.org/10.1111/pcn.12544

Takahashi K, Mizuno K, Sasaki AT, Wada Y, Tanaka M, Ishii A, Tajima K, Tsuyuguchi N, Watanabe K, Zeki S, Watanabe Y (2015) Imaging the passionate stage of romantic love by dopamine dynamics. Front Hum Neurosci 9:191. https://doi.org/10.3389/fnhum.2015.00191

Tata JR (2005) One hundred years of hormones. EMBO Rep 6 (6):490–496. https://doi.org/10.1038/sj.embor.7400444
Thomas CM, Nielsen KM (2005) Mechanisms of, and barriers to, horizontal gene transfer between bacteria. Nat Rev. Microbiol 3 (9):711–721. https://doi.org/10.1038/nrmicro1234
Tiihonen J, Rasanen P, Hakko H (1997) Seasonal variation in the occurrence of homicide in Finland. Am J Psychiatry 154 (12):1711–1714. https://doi.org/10.1176/ajp.154.12.1711
Todd EV, Liu H, Muncaster S, Gemmell NJ (2016) Bending Genders: The Biology of Natural Sex Change in Fish. Sex Dev 10 (5–6):223–241. https://doi.org/10.1159/000449297
Toro CA, Zhang L, Cao J, Cai D (2019) Sex differences in Alzheimer's disease: Understanding the molecular impact. Brain Res 1719:194–207. https://doi.org/10.1016/j.brainres.2019.05.031
Uddin LQ, Nomi JS, Hebert-Seropian B, Ghaziri J, Boucher O (2017) Structure and Function of the Human Insula. J Clin Neurophysiol 34 (4):300–306. https://doi.org/10.1097/WNP.0000000000000377
Uhl G, Zimmer SM, Renner D, Schneider JM (2015) Exploiting a moment of weakness: male spiders escape sexual cannibalism by copulating with moulting females. Sci Rep 5:16928. https://doi.org/10.1038/srep16928
Van Haastert PJ, Devreotes PN (2004) Chemotaxis: signalling the way forward. Nat Rev. Mol Cell Biol 5 (8):626–634. https://doi.org/10.1038/nrm1435
Vasey PL, Leca JB, Gunst N, VanderLaan DP (2014) Female homosexual behavior and intersexual mate competition in Japanese macaques: possible implications for sexual selection theory. Neurosci Biobehav Rev 46 Pt 4:573–578. https://doi.org/10.1016/j.neubiorev.2014.09.002
Vazquez MJ, Velasco I, Tena-Sempere M (2019) Novel mechanisms for the metabolic control of puberty: implications for pubertal alterations in early-onset obesity and malnutrition. J Endocrinol 242 (2):R51-R65. https://doi.org/10.1530/JOE-19-0223
Veenema AH (2012) Toward understanding how early-life social experiences alter oxytocin- and vasopressin-regulated social behaviors. Horm Behav 61 (3):304–312. https://doi.org/10.1016/j.yhbeh.2011.12.002
Ventura-Clapier R, Dworatzek E, Seeland U, Kararigas G, Arnal JF, Brunelleschi S, Carpenter TC, Erdmann J, Franconi F, Giannetta E, Glezerman M, Hofmann SM, Junien C, Katai M, Kublickiene K, Konig IR, Majdic G, Malorni W, Mieth C, Miller VM, Reynolds RM, Shimokawa H, Tannenbaum C, D'Ursi AM, Regitz-Zagrosek V (2017) Sex in basic research: concepts in the cardiovascular field. Cardiovasc Res 113 (7):711–724. https://doi.org/10.1093/cvr/cvx066
Vijayakumar N, Op de Macks Z, Shirtcliff EA, Pfeifer JH (2018) Puberty and the human brain: Insights into adolescent development. Neurosci Biobehav Rev 92:417–436. https://doi.org/10.1016/j.neubiorev.2018.06.004
Vladimirov N, Sourjik V (2009) Chemotaxis: how bacteria use memory. Biol Chem 390 (11):1097–1104. https://doi.org/10.1515/BC.2009.130
Walum H, Westberg L, Henningsson S, Neiderhiser JM, Reiss D, Igl W, Ganiban JM, Spotts EL, Pedersen NL, Eriksson E, Lichtenstein P (2008) Genetic variation in the vasopressin receptor 1a gene (AVPR1A) associates with pair-bonding behavior in humans. Proc Natl Acad Sci U S A 105 (37):14153–14156. https://doi.org/10.1073/pnas.0803081105
Wang Y, Chen CL, Iijima M (2011) Signaling mechanisms for chemotaxis. Dev Growth Differ 53 (4):495–502. https://doi.org/10.1111/j.1440-169X.2011.01265.x
White A, Levine R (1982) History of Hormones. In: Goldberger RF, Yamamoto KR (eds) Biological Regulation and Development: Hormone Action. Springer US, Boston, MA, pp. 1–24. https://doi.org/10.1007/978-1-4684-1125-6_1
Witt M, Wozniak W (2006) Structure and function of the vomeronasal organ. Adv Otorhinolaryngol 63:70–83. https://doi.org/10.1159/000093751
Wood CL, Lane LC, Cheetham T (2019) Puberty: Normal physiology (brief overview). Best Pract Res Clin Endocrinol Metab 33 (3):101265. https://doi.org/10.1016/j.beem.2019.03.001

Wroblewski EE, Murray CM, Keele BF, Schumacher-Stankey JC, Hahn BH, Pusey AE (2009) Male dominance rank and reproductive success in chimpanzees, Pan troglodytes schweinfurthii. Anim Behav 77 (4):873–885. https://doi.org/10.1016/j.anbehav.2008.12.014

Xiao YH, Zunic-Kosi A, Zhang LW, Prentice TR, McElfresh JS, Chinta SP, Zou YF, Millar JG (2015) Male adaptations to minimize sexual cannibalism during reproduction in the funnel-web spider Hololena curta. Insect Sci 22 (6):840–852. https://doi.org/10.1111/1744-7917.12243

Young LJ (2003) The Neural Basis of Pair Bonding in a Monogamous Species: A Model for Understanding the Biological Basis of Human Behavior. In: Wachter KW, Bulatao RA (eds) National Research Council (US) Panel for the Workshop on the Biodemography of Fertility and Family Behavior. National Academies Press, Washington,

Zeki S (2007) The neurobiology of love. FEBS Lett 581 (14):2575–2579. https://doi.org/10.1016/j.febslet.2007.03.094

Ziegler TE, Prudom SL, Zahed SR (2009) Variations in male parenting behavior and physiology in the common marmoset. Am J Hum Biol 21 (6):739–744. https://doi.org/10.1002/ajhb.20920

Zietsch BP, Westberg L, Santtila P, Jern P (2015) Genetic analysis of human extrapair mating: heritability, between-sex correlation, and receptor genes for vasopressin and oxytocin. Evolution and Human Behavior 36 (2):130–136. https://doi.org/10.1016/j.evolhumbehav.2014.10.001

Zuk M (2016) Mates with Benefits: When and How Sexual Cannibalism Is Adaptive. Curr Biol 26 (23):R1230-R1232. https://doi.org/10.1016/j.cub.2016.10.017

# Index





G. Majdic, *Soul Mate Biology*, Fascinating Life Sciences,
https://doi.org/10.1007/978-3-030-67212-6

The manufacturer's authorised representative in the EU is Springer Nature Customer Service Centre GmbH, Europaplatz 3, 69115 Heidelberg, Germany. If you have any concerns regarding our products, please contact ProductSafety@springernature.com

Printed and bound by CPI Group (UK) Ltd, Croydon, CR0 4YY
30/09/2026
02225690-0006